Lecture Notes in Mathematics

345

Proceedings of a Conference on Operator Theory

Dalhousie University, Halifax, Nova Scotia
April 13th and 14th, 1973

Edited by P. A. Fillmore
Dalhousie University, Halifax, N.S./Canada

Springer-Verlag
Berlin · Heidelberg · New York 1973

AMS Subject Classifications (1970): 46-02, 46L05, 47-02, 47A15, 47A55, 47B05, 47B10, 47B3
47B47, 47C10

ISBN 3-540-06496-6 Springer-Verlag Berlin · Heidelberg · New York
ISBN 0-387-06496-6 Springer-Verlag New York · Heidelberg · Berlin

Offsetdruck: Julius Beltz, Hemsbach/Bergstr.

PREFACE

During the past year a number of significant advances have been made in the study of Hilbert space operators with compact self-commutators. An informal mini-conference was desirable in order to explore further the close relationships among the results of the several groups and individuals working in this area. Such a conference was held at Dalhousie University, Halifax, N. S., on April 13th and 14th, 1973. This volume contains the Proceedings of that conference; it is presented in the hope of making the results available to the general mathematical public as quickly and in as convenient a form as possible.

We wish to thank the staff of Dalhousie University for help in organizing the conference and preparing this volume, Paula Flemming and Gretchen Smith for typing the manuscript, the National Science Foundation (U.S.A.) and the National Research Council (Canada) for financial support, and Springer-Verlag for prompt and efficient publication of the Proceedings.

Peter Fillmore
Editor

CONTENTS

1. C. A. BERGER and B. I. SHAW

Intertwining, analytic structure, and the trace norm estimate 1

2. L. A. COBURN

Toeplitz operators on odd spheres 7

3. R. G. DOUGLAS and CARL PEARCY

Invariant subspaces of non-quasitriangular operators 13

4. L. G. BROWN, R. G. DOUGLAS and P. A. FILLMORE

Unitary equivalence modulo the compact operators and extensions of C*-algebras 58

5. JEROME KAMINKER and CLAUDE SCHOCHET

Ext(X) from a homological point of view 129

6. J. WILLIAM HELTON and ROGER E. HOWE

Integral operators: commutators, traces, index, and homology 141

7. L. G. BROWN

The determinant invariant for operators with trace class self commutators 210

PARTICIPANTS

C. A. Berger

L. G. Brown

L. A. Coburn

J. A. Deddens

R. G. Douglas

M. Edelstein

P. A. Fillmore

J. W. Helton

R. E. Howe

J. Kaminker

E. Nordgren

S. K. Parrott

C. M. Pearcy

C. Schochet

B. I. Shaw

J. G. Stampfli

S. Swaminathan

INTERTWINING, ANALYTIC STRUCTURE, AND THE TRACE NORM ESTIMATE

C.A. Berger and B.I. Shaw

The authors have shown that if A is a k-rationally multicyclic hyponormal operator on a Hilbert space H, then $[A^*,A]$ lies in trace class, with trace norm bounded by k/π Area (sp(A)) [1]. This estimate is exact for certain "analytic models". It would be enlightening to focus on the means by which it is transferred to the operator A. The notation and definitions in this paper will conform to [1].

Let us first establish the result that is to be transferred. $(1/2\pi)d\theta$ is the normalized Haar measure on the unit circle ∂D. The operator S is the unilateral shift on $H = H^2(\partial D,(1/2\pi)d\theta)$, acting as multiplication by z.

Theorem 1. If $H = H^2(\partial D,(1/2\pi)d\theta)$, $f \in H^\infty(\partial D,(1/2\pi)d\theta)$, then $\operatorname{tr}[T_f^*, T_f] = 1/\pi \int_D |f'|^2 dA$, where dA is planar Lesbegue measure.

Proof. Let f be represented by the power series $f(z) = \sum_{m=0}^{\infty} c_m z^m$. For $n = 0,1,..$ let $e_n = e^{inx}$. When written with respect to the complete orthonormal basis $\{e_n\}_{n=0}^{\infty}$,

$$\begin{aligned}\operatorname{tr}[T_f^*, T_f] &= \sum_{n=0}^{\infty} (||T_f e_n||^2 - ||T_f^* e_n||^2) \\ &= \sum_{n=0}^{\infty} (||\sum_{m=0}^{\infty} c_m S^m e_n||^2 - ||\sum_{m=0}^{\infty} \bar{c}_m S^{*m} e_n||^2) \\ &= \sum_{n=0}^{\infty} (||\sum_{m=1}^{\infty} c_m e_{n+m}||^2 - ||\sum_{m=1}^{n} \bar{c}_m e_{n-m}||^2) .\end{aligned}$$

The above sum may be rewritten as

$$\sum_{n=0}^{\infty} (\sum_{m=n+1}^{\infty} |c_m|^2) = \sum_{m=1}^{\infty} \sum_{n=0}^{m-1} |c_m|^2 = \sum_{m=1}^{\infty} m|c_m|^2 .$$

By the classical "area formula", $\sum_{m=1}^{\infty} m|c_m|^2 = 1/\pi \int_D |f'|^2 dA$.

This proves the theorem.

Remark: If T is any subnormal weighted shift of norm 1 , then $tr[f(T)^*, f(T)] = 1/\pi \int_D |f'|^2 dA$. The computation is also direct.

Corollary 1. If U is an open bounded simply connected set in the plane and $H = A^2(U, dA)$ (the square integrable analytic functions on U), $f \varepsilon A^\infty(U, dA)$, then

$$tr[T_f^*, T_f] = 1/\pi \int_U |f'|^2 dA .$$

This result is true whether or not U is simply connected, but the stronger result requires more machinery [1].

Operators of the form T_z on $A^2(U, dA)$, or direct sums of such may be used to provide "analytic models" for any operator.

For F a compact set in the plane $R(F)$ will denote the rational functions with poles off F .

Definition: An operator $T \varepsilon L(H)$ is said to be an "elementary universal model" on the bounded open set G if, whenever $A \varepsilon L(K)$ $(\dim(K) = \aleph_0)$ satisfies

1. $sp(A) \subset G$

2. There exists an $x_o \varepsilon K$ such that the closure of $\bigvee\{r(A)x_o : r \varepsilon R(\bar{G})\} = K$,

then there exists $W \varepsilon L(H,K)$ such that

a. $\overline{R(W)} = K$, $N(W) = \{0\}$

b. $WT = AW$

Note that $W(N(T-\lambda I)) \subset N(A-\lambda I)$ for all λ . Thus if T is universal, $N(T-\lambda I) = \{0\}$ for all λ . Furthermore since $W^*A^* = T^*W^*$, it follows that $W^*N(A^*-\bar{\lambda}I) \subset N(T^*-\lambda I)$.

Suppose U is an open set with Jordan boundary whose closure is contained in the open set G . Take $A = T_z$ on $R^2(\bar{U}, dA)$, k_λ the reproducing kernel at λ . Then for some W as above $T^*W^*k_\lambda = \bar{\lambda}W^*k_\lambda$. Thus $\{W^*k_\lambda : \lambda \varepsilon U\}$ is a strongly continuous conjugate analytic vector field with $T^*(W^*k_\lambda) = \bar{\lambda}W^*k_\lambda$.

<u>Theorem 2.</u> If $\{k_\lambda:\lambda\varepsilon G\}\subset H$ is a strongly continuous conjugate analytic vector field on the open set G and $T\varepsilon L(H)$ satisfies $T^*k_\lambda = \bar\lambda k_\lambda$, and $A\varepsilon L(K)$, $sp(A)\subset G$, and the closure of $\vee\{r(A)x_o : r\varepsilon R(\bar G)\} = K$, then there exists $W\varepsilon L(H,K)$ such that

1. $N(W) = N(W^*) = \{0\}$
2. $WT = AW$
3. W is in trace class.
4. If G', an open set, satisfies $sp(A)\subset G'\subset\overline{G'}\subset G$, then $||W|| \le ||W||_2 \le C_{G'} \sup_{\lambda\varepsilon G'} ||k_\lambda|| \sup_{\lambda\varepsilon G'} ||R_\lambda(A)||\ ||x_o||$, where $||\ ||_2$ denotes the Hilbert-Schmidt norm.

<u>Proof</u>. As in [1], for any $u\varepsilon H$, $\hat u(z) = (u,k_z)$, $z\varepsilon G$, is an analytic function, and we may define

$$\hat u(A) = -1/2\pi i \int_{\Gamma'} (u,k_z)R_z(A)d\gamma(z)$$

for Γ' a finite union of smooth Jordan curves in G bounding G' from G^c. We define W by $Wu = \hat u(A)x_o$, for $u\varepsilon H$. We observe that for all $u\varepsilon H$, $(Tu,k_z) = (u,T^*k_z) = (u,\bar z k_z) = z(u,k_z)$, or $\widehat{Tu} = z\,\hat u$. Therefore for all $u\varepsilon H$, $WTu = \widehat{Tu}(A)x_o = \widehat{zu}(A)x_o$ $= A\hat u(A)x_o = AWu$ and $WT = AW$, which proves 2.

We will now show that W is in trace class. Let Γ_1 be a finite set of smooth Jordan curves bounding $sp(A)$ from G'^c, and let Γ_2 be another such set bounding Γ_1 from G'^c, and Γ_3 a third bounding Γ_2 from G'^c. Let λ_i be arc length measure on Γ_i. Let H_i be the closure of the functions $\{\hat u:u\varepsilon H\}$ in $L^2(\lambda_i)$. Let $W_3:H\to H_3$ be defined by $W_3u = \hat u|\Gamma_1$. Observe that $T_i = T_z|H_i$ admits an analytic evaluation in the sense of [1]. Define $W_i\,u = \hat u|\Gamma_i$ for $u\varepsilon H_{i+1}$ for $i = 1,2$ and $W_o\,u = \hat u(A)x_o$ for $u\varepsilon H_1$. We have $W = W_0W_1W_2W_3$. Each W_i is bounded since $z\to k_z$ is strongly continuous and conjugate analytic on G. It is easy to represent W_2 and W_1 as integral operators with square summable kernels. These kernels are just the usual Szëgo kernels. Thus W_2 and W_1 are Hilbert-Schmidt operators and so W_1W_2 , being a product of Hilbert-Schmidt operators, is a trace class operator. This proves 3.

To prove 1., suppose $u \in N(W)$ or $\hat{u}(A)x_o = 0$. The analytic function $\hat{u}$ may be factored as fg where f does not vanish on $\overline{G'}$ and g is a polynomial with at most a finite number of zeros on $\overline{G'}$. Clearly $\hat{u}(A) = f(A)g(A)$. By the spectral mapping theorem as applied to the Riesz functional calculus, $f(A)$ is invertible. Thus $\hat{u}(A)x_o = 0$ implies that $g(A)x_o = 0$. However, x_o is a cyclic vector. Thus $sp(A)$ consists only of a finite number of points contained in $g(0)^{-1}$. Any rational function with poles off $sp(A)$ may be approximated uniformly by polynomials on $sp(A)$. Hence A has a cyclic vector in the usual sense and satisfies a polynomial equation, which contradicts $\dim(K) = \aleph_o$. By construction W has dense range; thus $N(W^*) = \{0\}$, and 1. is proven.

The proof of 4. is a direct estimate from the Riesz integral. We may write $Wu = \langle u, -1/2\pi i \int_\Gamma k_z \otimes R_z(A)x_o d\gamma \rangle$, where $\otimes$ denotes the tensor product between H and K . Hence we see that W is a convolution operator. This proves the theorem.

If an operator is k-rationally cyclic, intertwinings may be constructed from direct sums of elementary universal models. For example, let $D' = \{z : |z| < 1 \ , z \neq it \ , 0 \leq t \leq 1\}$, $K = A^2(D', dA)$, $A = T_z | A^2(D', dA)$. A is a rationally cyclic subnormal operator on K with two independent generators, the vectors 1 and $\sqrt{z}$. That A does not have a single cyclic vector follows from an unpublished result of D.J. Newman.

If $H = H^2(D, dA) \oplus H^2(D, dA)$ (D is the unit disc), then W may be defined on the two free generators of H , $(0,1)$, $(1,0)$, by $W(0,1) = \sqrt{z}$ and $W(1,0) = 1$. By introducing more slits in D we can produce operators requiring more than two generators, indeed, requiring infinitely many.

These intertwinings do transfer some residuum of analytic

structure to the covered operator. We note that elementary universal models can never have semi-Fredholm index which is positive, but rather look like shifts. The hyponormal operators T_z described earlier are of course elementary universal models.

The following theorem verifies one's hope that the analytic structure of the model may be transferred to the covered operator by an intertwining map.

Theorem 3. Let $T \varepsilon \mathcal{L}(H)$, $A \varepsilon \mathcal{L}(K)$, both T and A be hyponormal operators, and $W \varepsilon L(H,K)$ such that

1. $WT = AW$
2. W is a Hilbert-Schmidt operator.
3. $\overline{\mathcal{R}(W)} = K$.

If $\mathcal{N}(W) = N$, then

$$\operatorname{tr}[T^*,T] \geq \operatorname{tr}[(T|_N)^*, (T|_N)] + \operatorname{tr}[A^*,A]$$

(Note that N is an invariant subspace for T and $(T|_N)$ is hyponormal).

The proof of this theorem is lengthy. A proof in the case where W lies in trace class is scheduled to appear in [1]. The method results in the following theorem:

Theorem 4. (Main Theorem) If A is a k-multicyclic hyponormal operator, then $[A^*,A]$ is in trace class, and $\operatorname{tr}[A^*,A] \leq (K/\pi)\omega(\operatorname{sp}(A))$, where ω is planar Lesbegue measure.

Corollary. (Putnam's Theorem). If $A \varepsilon \mathcal{L}(H)$ is hyponormal, then $\|[A^*,A]\| \leq (1/\pi)\omega(\operatorname{sp}(A))$.

R. Douglas and C. Pearcy have been able to use the method of Theorem 2 to simplify the Foias theorem: "Invariant subspaces Of Non-Quasitriangular Operators" [2]. (See also [3]). Specifically, they observed that an absolutely continuous normal operator on Jordan arcs is the direct sum of a universal model and another operator.

Thus, they were able to transfer certain weak localization properties of the normal operator to the covered operator.

<u>References</u>

1. C.A. Berger and B.I. Shaw, Self-commutators of multicyclic hyponormal operators are always trace class, to appear in Bull. Amer. Math. Soc. (1973).

2. R.G. Douglas and C. Pearcy, Invariant subspaces of non-quasitriangular operators, these Notes.

3. C. Foias, C. Apostol, and D. Voiculescu, Some results on non-quasitriangular operators, II, III, IV, Rev. Roum. Math. Pures et Appl., (1973).

Yeshiva University
Belfer School of Science
New York, N.Y.

The City University
Bernard M. Baruch College
New York, N.Y.

TOEPLITZ OPERATORS ON ODD SPHERES*

L. A. Coburn

In this note I describe the structure of the C^*-algebras generated by certain Toeplitz operators on Hardy spaces of spheres in several complex variables. The proof of the main result is completely self-contained and uses only relatively elementary facts about the Hardy spaces involved and about C^*-algebras. As a direct corollary of this result, I obtain the classical and somewhat non-trivial fact that if ϕ is a function analytic on the open unit ball B^{2n} in C^n and continuous on its boundary, the unit sphere S^{2n-1}, then for $n > 1$, $\phi(B^{2n})^- = \phi(S^{2n-1})$ (here U^- denotes the closure of the set U). For a more complete discussion of the algebras considered here, the reader is referred to [1].

In what follows, we let $L^2(S^{2n-1})$ be the usual space of complex-valued functions on S^{2n-1} which are square-integrable with respect to the usual surface area measure. We also consider the Hardy subspace $H^2(S^{2n-1})$ consisting of functions in $L^2(S^{2n-1})$ which are radial boundary-values of functions analytic in B^{2n}. For the basic properties of $H^2(S^{2n-1})$ including the fact that functions in $H^2(S^{2n-1})$ have unique analytic extensions to B^{2n}, see [3]. Let P_n be the orthogonal projection from $L^2(S^{2n-1})$ onto $H^2(S^{2n-1})$. For ϕ a bounded measurable complex-valued function on S^{2n-1} we define the Toepliz operator T_ϕ on $H^2(S^{2n-1})$ by

$$T_\phi f = P_n(\phi \cdot f) .$$

The main result of this note describes the structure of the C^*-algebra $\tau(S^{2n-1})$ generated by all Toeplitz operators on $H^2(S^{2n-1})$ associated with ϕ in $C(S^{2n-1})$, the sup norm algebra of all complex-valued continuous functions on S^{2n-1}.

*Research supported by grants of the National Science Foundation.

We will make use of many easily checked results such as the linearity of the map $\phi \to T_\phi$ as well as the obvious norm estimate $||T_\phi|| \leq ||\phi||_\infty$ and the fact that $T_\phi^* = T_{\bar{\phi}}$ (here $\bar{\phi}$ denotes the complex conjugate of ϕ). Moreover, if ψ is bounded and in $H^2(S^{2n-1})$ and ϕ is merely bounded, we have $T_\phi T_\psi = T_{\phi\psi}$. We will also use the natural orthonormal basis for $H^2(S^{2n-1})$ given by

$$e_k = \frac{1}{\sqrt{2\pi^n}} \sqrt{\frac{(n+|k|-1)!}{k!}}\, z^k$$

where $k = (k_1, k_2,\dots,k_n)$ is an n-tuple of non-negative integers and we take $|k| \equiv k_1 +\dots+ k_n$, $k! \equiv k_1!\, k_2!\, k_3! \dots k_n!$, $z^k \equiv z_1^{k_1} z_2^{k_2} \dots z_n^{k_n}$ where $z = (z_1, z_2,\dots,z_n)$ is a point in C^n. The reproducing kernel for $H^2(S^{2n-1})$ is given (see [3]) by

$$K(\lambda,z) = \frac{(n-1)!}{2\pi^n}(1-\bar{\lambda}.z)^{-n}$$

where $\bar{\lambda} \equiv (\bar{\lambda}_1, \bar{\lambda}_2,\dots,\bar{\lambda}_n)$ and

$$\bar{\lambda}\cdot z = \bar{\lambda}_1 z_1 +\dots+ \bar{\lambda}_n z_n .$$

The structure of $\tau(S^{2n-1})$ is given by the following two Lemmas and the Theorem.

<u>Lemma 1</u>. The algebra $\tau(S^{2n-1})$ is irreducible.

<u>Proof</u>. Assume there is a projection P which commutes with all the T_ϕ where ϕ are continuous on S^{2n-1} and extend analytically to B^{2n}. The set of all such ϕ is dense in $H^2(S^{2n-1})$ (polynomials in the z_i will do). Now let $P1 = g$. Then $PT_\phi 1 = T_\phi P1$ so $P\phi = \phi g$ for a dense set of ϕ's in $H^2(S^{2n-1})$. A standard argument now shows that $Pf = fg$ for <u>all</u> f in $H^2(S^{2n-1})$ and that $|g(z)| \leq 1$ a.e. so g is the boundary-value function of a bounded analytic function in B^{2n} (also designated by g) and $P = T_g$. Now for $|\lambda| < 1$ (i.e. λ in B^{2n}) standard properties of the reproducing kernel yield

$$T_g^* K(\lambda,\cdot) = \overline{g(\lambda)}\, K(\lambda,\cdot) .$$

Since $P = T_g$ is a projection, its spectrum must be contained in $\{0,1\}$ and it follows that $g(B^{2n}) \subset \{0,1\}$. Hence $g \equiv 0$ or $g \equiv 1$ and $P = 0$ or $P = I$.

Lemma 2. If T_ϕ is compact then $\phi \equiv 0$ for ϕ in $C(S^{2n-1})$.

Proof. Suppose there is an x_o in S^{2n-1} with $\phi(x_o) \neq 0$. For a sequence of λ_m in B^{2n} with $\lambda_m \to x_o$ we consider the functions in $H^2(S^{2n-1})$

$$f_m(z) = K(\lambda_m,z)[K(\lambda_m,\lambda_m)]^{-\frac{1}{2}} .$$

It is an easy computation that $||f_m|| = 1$ and the f_m converge uniformly to zero in the complement of any open neighborhood of x_o in S^{2n-1}. It follows that $||T_\phi f_m - \phi(x_o) f_m|| \to 0$. Now we can assume without loss of generality that $f_m \rightharpoonup g$ (weakly). If T_ϕ is compact then $T_\phi f_m \to T_\phi g$ and so we have $f_m \to g$ (in norm). Thus $||g|| = 1$ and f_m converges to g in measure. But f_m converges to zero pointwise a.e. so $g = 0$ a.e., a contradiction.

In what follows, we will use the auxilliary operator

$$S\, e_k \equiv \begin{cases} e_{k_1+1,k_2-1,k_3,\ldots k_n} & k_2 \neq 0 \\ 0 & k_2 = 0 \end{cases}$$

We can now prove a result first established in [1].

Theorem. The algebra $\tau(S^{2n-1})$ contains the full algebra of compact operators $\mathcal{K}$. Moreover, $\tau(S^{2n-1}) = \{T_\phi + K$ for all ϕ in $C(S^{2n-1})$ and K in $\mathcal{K}\}$ and this representation is unique. The quotient $\tau(S^{2n-1})/\mathcal{K}$ is naturally identified with $C(S^{2n-1})$ by the map $\sigma(T_\phi + K) = \phi$.

Proof. We first note that by direct computation

$$(T_{z_1}^* T_{z_1} - T_{z_1} T_{z_1}^*)e_k = \begin{cases} (n+|k|)^{-1} e_k , & k_1=0 \\ (n+|k|)^{-1}[1-k_1(n+|k|-1)^{-1}]e_k , & k_1\neq 0 \end{cases}$$

and for $n > 1$

$$T_{z_1} T_{z_2}^* - T_{z_2}^* T_{z_1} = SD$$

where

$$D e_k = (n+|k|)^{-1} [k_2(1+k_1)]^{\frac{1}{2}} (n+|k|-1)^{-1} e_k .$$

Now D and $T_{z_1}^* T_{z_1} - T_{z_1} T_{z_1}^*$ are both diagonal and positive with k-th term going to zero as $|k|$ increases. It follows that these operators are compact. Uisng approximation by polynomials in z and $\bar{z}$ and symmetry among the z_i as well as the elementary properties of the T_ϕ previously mentioned, it is now easy to see that $T_{\phi_1} T_{\phi_2} - T_{\phi_1 \phi_2}$ is compact for all ϕ_i in $C(S^{2n-1})$. It follows from Lemma 1 and a well-known elementary result in C^*-algebras [2,p.85] that $\mathcal{K} \subset \tau(S^{2n-1})$.

Now for ϕ in $C(S^{2n-1})$ we let $[T_\phi]$ be the class of T_ϕ in $\tau(S^{2n-1})/\mathcal{K}$. The map $\phi \to [T_\phi]$ is easily seen to be a *-homomorphism from $C(S^{2n-1})$ into $\tau(S^{2n-1})/\mathcal{K}$ with range $Q = \{[T_\phi]: \phi \in C(S^{2n-1})\}$ so by another standard result [2,p.18] Q must be a closed *-subalgebra of $\tau(S^{2n-1})/\mathcal{K}$. But Q contains the quotient classes of all generators of $\tau(S^{2n-1})$ so $Q = \tau(S^{2n-1})/\mathcal{K}$ and it follows that

$$\tau(S^{2n-1}) = \{T_\phi + K : \phi \in C(S^{2n-1}) \text{ and } K \in \mathcal{K}\} .$$

Finally, by Lemma 2 the map $\phi \to [T_\phi]$ is an isomorphism and the representation of $\tau(S^{2n-1})$ as sums $T_\phi + K$ is unique.

We recall that an operator A is called <u>Fredholm</u> if A has closed range and kernel A and kernel A^* are both finite dimensional. For a discussion of the basic properties of such operators see [4]. We can now state

<u>Corollary 1</u>. For ϕ in $C(S^{2n-1})$, T_ϕ is Fredholm if and only if $\phi \neq 0$ in S^{2n-1}. Moreover, for $n > 1$ the index of Fredholm T_ϕ must be zero (i.e. the kernels of T_ϕ and T_ϕ^* must have the <u>same</u>

finite dimension).

Proof. The first statement follows from Atkinson's theorem [4,p.120] and our main result. The second statement follows from the continuity of the index [4,p.122] and the fact that for $n > 1$ all non-vanishing complex-valued continuous functions on S^{2n-1} are homotopic to the constant function 1. The latter fact is from elementary topology (it is the same as stating $\pi_{2n-1}(S^1) = 0$ for $n>1$).

I can now give the promised application to analytic function theory in several variables.

Corollary 2. For $n > 1$, let ϕ be analytic on B^{2n} and continuous on the boundary S^{2n-1}. Then $\phi(B^{2n})^- = \phi(S^{2n-1})$.

Proof. First we recall that $T_\phi^* K(\lambda,\cdot) = \overline{\phi(\lambda)}\, K(\lambda,\cdot)$ for λ in B^{2n} so that $\phi(B^{2n}) \subset$ spectrum (T_ϕ). It now suffices to show that spectrum $(T_\phi) \subset \phi(S^{2n-1})$. To do this we need only show that $\phi \neq 0$ on S^{2n-1} implies that T_ϕ is invertible. Now by Corollary 1, $\phi \neq 0$ on S^{2n-1} implies that T_ϕ is Fredholm with

$$\text{dimension kernel } (T_\phi) = \text{dimension kernel } (T_\phi^*) .$$

Further, $T_\phi f = \phi f$ so kernel $(T_\phi) = \{0\}$ and it follows that T_ϕ is invertible.

I conclude with a question which is suggested by the main result. It is an easy consequence of the main result that for ϕ continuous on S^{2n-1}, $||T_\phi|| = ||\phi||_\infty$. Does this equality also hold for ϕ merely bounded measurable? For $n = 1$ and Toeplitz operators in a variety of other settings the equality is known to hold for all bounded measurable ϕ, but the proofs which are known (there are several) all seem to use properties of the Hardy spaces which are either not evident or not true for $H^2(S^{2n-1})$, $n > 1$. The answer to this question should also shed some light on other interesting questions in the case $n > 1$ which arise by looking at the vast literature on Toeplitz operators in the case $n = 1$.

References

[1] Coburn, L. A., Singular integral operators and Toeplitz operators on odd spheres, Indiana J., (1973).

[2] Diximier, J., Les C^*-algebres et leurs representations, (1969).

[3] Koranyi, A., The Poisson integral for generalized half-planes and bounded symmetric domains, Annals. Math. 82(1965) 335-350.

[4] Palais, R. S., Seminar on the Atiyah-Singer index theorem, Annals of Math. Studies 57, (1965).

Yeshiva University
Belfer Graduate School
New York, N.Y.

INVARIANT SUBSPACES OF NON-QUASITRIANGULAR OPERATORS

R. G. Douglas and Carl Pearcy

1. Introduction

In this paper we shall be concerned with separable, infinite dimensional, complex Hilbert spaces, and all operators on such spaces under discussion will be assumed to be bounded and linear. If H is such a space, the algebra of all operators on H will be denoted by $L(H)$. An operator T on H is said to be quasitriangular if there exists a sequence $\{P_n\}$ of (orthogonal) projections of finite rank on H converging strongly to 1_H and satisfying

$$||P_n T P_n - T P_n|| \to 0 .$$

The class of quasitriangular operators was introduced in [14] by Halmos, who recognized the pertinence of the concept to invariant subspace problems. In this pioneering paper [14], various alternate characterizations of the class of quasitriangular operators were given, various important classes of operators were shown to be quasitriangular, and the existence of non-quasitriangular operators was established. Subsequently, quasitriangular operators were studied in [1], [10], [13] and [16], and in [13] the following result was obtained: if $T \in L(H)$ and there exists a complex number λ such that $T-\lambda$ is a semi-Fredholm operator with negative index, then T is non-quasitriangular. However, the class of non-quasitriangular operators remained relatively mysterious until very recently, when Apostol, Foiaş, Voiculescu and Zsido studied non-quasitriangular operators in a sequence of four papers [2]-[5]. This study culminated in the proof in [5] of the following beautiful converse of the result just quoted from [13].

THEOREM A (Apostol, Foiaş, Voiculescu). If T is any non-quasitriangular operator on H, then there exists a complex number λ such that $T-\lambda$ is a semi-Fredholm operator with negative index.

An immediate corollary of this theorem and the well-known theory of Fredholm operators is the affirmative answer to an earlier conjecture of Pearcy (cf. [15], Problem X), namely:

COROLLARY B. The adjoint of every non-quasitriangular operator has nonempty point spectrum.

Of course, a trivial corollary of Corollary B is the following remarkable fact.

COROLLARY C. Every non-quasitriangular operator has a non-trivial invariant (in fact, hyperinvariant) subspace.

The purpose of this paper is to present a proof of Theorem A that is somewhat different and, hopefully, more perspicuous than the original proof of Theorem A in [5]. We have tried to make this paper essentially self-contained, and we use neither the theory of decomposable operators nor the function q introduced in [1]. Many of the important new ideas that appear herein, however, are due to the authors of [5], and we try to point these out during the course of the proof.

The present authors gratefully acknowledge to Apostol, Foias and Voiculescu the early receipt of a preprint of [5].

2. Preliminaries

In this section we introduce some terminology and notation that will be needed in what follows.

The ideal of compact operators in $L(H)$ will be denoted by $\mathbb{K}$, and the canonical homomorphism of $L(H)$ onto the Calkin algebra $L(H)/\mathbb{K}$ will be denoted by π. The spectrum of an operator T in $L(H)$ will be denoted by $\sigma(T)$, and the essential spectrum of T, i.e., the spectrum of $\pi(T)$ in $L(H)/\mathbb{K}$, will be denoted by $E(T)$.

The set of Fredholm operators in $L(H)$ will be denoted by (F), and the larger set of semi-Fredholm operators will be denoted by (SF). If $T \in (SF)$, the Fredholm index of T will be denoted by $i(T)$. We shall assume that the reader is familiar with the elementary theory of Fredholm and semi-Fredholm operators. In particular, one knows that $E(T) = \{\lambda \in \mathbb{C} : T-\lambda \notin (F)\}$. Also the <u>left essential spectrum</u> of T, denoted by $E_\ell(T)$ and defined by

$$E_\ell(T) = \{\lambda \in \mathbb{C} : \pi(T-\lambda) \text{ is not left invertible}\},$$

can be characterized as the set of all complex numbers λ such that either $T-\lambda$ has an infinite dimensional null space or $T-\lambda$ does not have closed range. Furthermore,

$$E(T)\backslash E_\ell(T) = \{\lambda \in \mathbb{C} : T-\lambda \in (SF) \text{ and } i(T-\lambda) = -\infty\}.$$

We shall also need the concept of the <u>Weyl spectrum</u> $W(T)$ of an operator T on H. The set $W(T)$ may be characterized as

$$\begin{aligned} W(T) &= \{\lambda \in \mathbb{C} : T-\lambda \notin (F), \text{ or } T-\lambda \in (F) \text{ and } i(T-\lambda) \neq 0\} \\ &= \bigcap_{K \in \mathbb{K}} \sigma(T+K). \end{aligned}$$

Of course, $E_\ell(T) \subset E(T) \subset W(T) \subset \sigma(T)$ for every operator T.

It will sometimes be convenient to use the notation introduced in this section for operators acting on separable Hilbert spaces other than H, and we shall do this without further comment.

3. The first reduction

In this section, we obtain the first important result (Theorem 3.2) along the road to a proof of Theorem A. We begin with a proposition which is a strengthening of [5, Theorem 2.2].

THEOREM 3.1 Let T belong to $L(H)$. Then T is unitarily equivalent to the sum of a compact operator on $H \oplus H$ and an operator on $\mathcal{H} \oplus \mathcal{H}$ of the form

$$\begin{pmatrix} N & A \\ 0 & S \end{pmatrix} \qquad (*)$$

where N is a diagonable normal operator of uniform infinite multiplicity such that $\sigma(N) = E_\ell(T)$, and where $\sigma(N) \subset E_\ell(S)$.

Proof. Fix a countable (finite or infinite) dense subset D of $E_\ell(T)$, and let $\{\lambda_n\}_{n=1}^\infty$ be a sequence such that each λ_n belongs to D and such that each number in D is repeated infinitely often in the sequence $\{\lambda_n\}$. Observe now that we can construct by induction an orthonormal sequence $\{e_n\}_{n=1}^\infty$ in $\mathcal{H}$ with the property that for each positive integer n , $||(T-\lambda_n)e_n|| < 1/2^n$. (This is possible because if $e_1,\ldots,e_k$ have been constructed with the desired properties, and $E = \vee\{e_1,\ldots,e_k\}$, then since $\lambda_{k+1} \in E_\ell(T)$, $T-\lambda_{k+1}$ cannot be bounded below on $E^\perp$. Thus there exists a unit vector e_{k+1} in $E^\perp$ such that $||(T-\lambda_{k+1})e_{k+1}|| < 1/2^{k+1}$.) Let $\mathcal{M} = \vee\{e_n\}_{n=1}^\infty$, and observe that we may assume that $\mathcal{H} \ominus \mathcal{M}$ is infinite dimensional by judiciously contributing infinitely many of the e_n to $\mathcal{H} \ominus \mathcal{M}$ if necessary. If N is the normal operator in $\mathcal{L}(\mathcal{M})$ defined by $Ne_n = \lambda_n e_n$ for each n , then $||Te_n - Ne_n|| < 1/2^n$, and it follows easily that if we identify each of $\mathcal{M}$ and $\mathcal{H} \ominus \mathcal{M}$ with a copy of $\mathcal{H}$ via a unitary isomorphism, then T becomes unitarily equivalent to an operator on $\mathcal{H} \oplus \mathcal{H}$ of the form

$$\begin{pmatrix} N+H_1 & A_1 \\ H_2 & S_1 \end{pmatrix}$$

where H_1 and H_2 are Hilbert-Schmidt operators. Clearly N has the desired properties, and we define K to be the compact operator on $\mathcal{H} \oplus \mathcal{H}$ given by the matrix

$$\begin{pmatrix} H_1 & 0 \\ H_2 & 0 \end{pmatrix}$$

Then T is unitarily equivalent to the sum of K and the operator

$$T_1 = \begin{pmatrix} N & A_1 \\ 0 & S_1 \end{pmatrix} .$$

Since N is unitarily equivalent to $N \oplus N$, clearly T_1 is unitarily equivalent to an operator on $\mathcal{H} \oplus \mathcal{H} \oplus \mathcal{H}$ of the form

$$\begin{pmatrix} N & 0 & A_{11} \\ 0 & N & A_{12} \\ 0 & 0 & S_1 \end{pmatrix} ,$$

and the result now follows by identifying $\mathcal{H} \oplus (\mathcal{H} \oplus \mathcal{H})$ with $H \oplus (\mathcal{H})$ and defining S to be the operator on $\mathcal{H}$ that corresponds under the identification of $\mathcal{H} \oplus \mathcal{H}$ with $\mathcal{H}$ to the 2×2 matrix

$$\begin{pmatrix} N & A_{12} \\ 0 & S_1 \end{pmatrix} .$$

It is an easy consequence of the techniques of §2 of [7] that the compact operator K in the above result may be taken to be a trace-class operator of arbitrarily small trace norm. Moreover, various results related to Theorem 3.1 have been recently obtained by several authors (cf. [8 , p. 69]).

For each T in $\mathcal{L}(\mathcal{H})$, we shall write $\omega(T)$ for the set of all complex numbers λ such that $T-\lambda \in (\mathcal{SF})$ and $-\infty \leq i(T-\lambda) < 0$. Observe that Theorem A is equivalent to the statement that if $T \in \mathcal{L}(\mathcal{H})$ and $\omega(T)$ is empty, then T is quasitriangular.

THEOREM 3.2 Suppose T' is an operator on $\mathcal{H}$ such that $\omega(T')$ is empty. Then T' is unitarily equivalent to the sum of a compact operator and an operator T on $\mathcal{H} \oplus \mathcal{H}$ of the form (*), where N is diagonable normal operator of uniform infinite multiplicity satisfying $\sigma(N) = E(T) = E(S)$, and where $\omega(S)$ is empty.

Proof. Since $\omega(T')$ is empty, $E(T') = E_{\ell}(T')$, and we apply

Theorem 3.1 to T' to obtain a unitary equivalence between T' and the sum of a compact operator K and an operator matrix T on $H \oplus H$ of the form (*), where N is a diagonal normal operator of uniform infinite multiplicity satisfying $\sigma(N) = E_\ell(T') = E(T') = E(T) \subset E_\ell(S) \subset E(S)$. To see that $E(T) = E(S)$ and that $\omega(S)$ is empty, suppose $\lambda \notin E(T)$. Then $T-\lambda$ is Fredholm and $i(T-\lambda) \geq 0$ (since $\omega(T)$ is empty). Furthermore, since $\sigma(N) = E(T)$, $N-\lambda$ is invertible. It now follows from easy matricial calculations that $\text{range}(T-\lambda) = H \oplus \text{range}(S-\lambda)$, that $\ker(T-\lambda)^* = (0) \oplus \ker(S-\lambda)^*$, and that $\dim[\ker(T-\lambda)] = \dim[\ker(S-\lambda)]$. Thus $S-\lambda \in (F)$ and $i(S-\lambda) = i(T-\lambda)$. Hence $\lambda \notin E(S)$, which proves that $E(T) = E_\ell(S) = E(S)$. Furthermore if λ is such that $(S-\lambda) \in (F)$, then $T-\lambda$ is Fredholm and $i(S-\lambda) = i(T-\lambda) \geq 0$. Since $E_\ell(S) = E(S)$,

$$\{\lambda \in \mathbb{C} : S-\lambda \in (SF) \text{ and } i(S-\lambda) = -\infty\} = \emptyset ,$$

and hence $\omega(S)$ is empty, so the proof is complete.

4. The second reduction

One knows from [14] that an operator T' is quasitriangular if and only if every operator that is unitarily equivalent to a compact perturbation of T' is quasitriangular. Using this fact and Theorem 3.2, one sees immediately that to prove Theorem A, it suffices to prove that every operator matrix T acting on $H \oplus H$ of the form (*) is quasitriangular, where N is a diagonable normal operator of uniform multiplicity such that $\sigma(N) = E(S)$, and where $\omega(S)$ is empty. In this section we present another important insight due to the authors of [5].

THEOREM 4.1 Let N be a diagonable normal operator in $L(H)$ of uniform infinite multiplicity, and let S be any operator in $L(H)$ such that $N \oplus S$ is quastitriangular. Then for every operator A

in $L(H)$, the 2×2 operator matrix (*) is quasitriangular.

Proof. If the spectrum of N is finite [infinite], then N is a finite [infinite] direct sum of operators each of which is scalar on an infinite dimensional subspace. In either case, there exists a sequence $\{V_n\}_{n=1}^{\infty}$ of isometries on H such that each V_n commutes with N , such that each V_n^* has finite dimensional kernel, and such that the sequence $\{V_n^*\}$ converges strongly to 0 . (In case $\sigma(N)$ is finite, there exists a pure isometry V of finite multiplicity that commutes with N , and we can take $V_n = V^n$. In case $\sigma(N)$ is infinite, identify N with an infinite direct sum of scalar operators each acting on one copy of H , let U be a shift of multiplicity one acting on H , and define $V_1 = U \oplus 1_H \oplus 1_H \oplus \ldots$, $V_2 = U^2 \oplus U \oplus 1_H \oplus \ldots$, etc.). Let S be an operator on H such that $N \oplus S$ is quasitriangular, and let $\{P_n\}$ be a sequence of projections of finite rank in $L(H \oplus H)$ such that $\{P_n\}$ converges strongly to $1_{H\oplus H}$ and such that $||(1-P_n)(N \oplus S)P_n|| \to 0$. Let A be an arbitrary operator in $L(H)$, and denote by $\tilde{A}$ the operator

$$\tilde{A} = \begin{pmatrix} 0 & A \\ 0 & 0 \end{pmatrix}$$

in $L(H \oplus H)$. Note that for each fixed positive integer n , the operator $\tilde{A}P_n$ is of finite rank, and thus, since the sequence $\{V_k^* \oplus 0\}_{k=1}^{\infty}$ converges strongly to 0 , there exists a positive integer $k_n \geq n$ such that

$$||(V_{k_n}^* \oplus 0)\tilde{A}P_n|| < 1/n .$$

(See, for example, Proposition 6.2 of [10].) Consider now the sequence $\{Q_n\}$ of finite rank projections in $L(H \oplus H)$ defined by

$$Q_n = [(1_H - V_{k_n}V_{k_n}^*) \oplus 0] + (V_{k_n} \oplus 1_H)P_n(V_{k_n}^* \oplus 1_H) .$$

It is easy to see that the sequence $\{Q_n\}$ converges strongly to

$1_{H\oplus H}$ (cf. [10], Corollary 2.2), and furthermore that the projection $1_{H\oplus H} - Q_n = 1-Q_n$ is given by

$$1-Q_n = (V_{k_n} \oplus 1_H)(1-P_n)(V_{k_n}^* \oplus 1_H) .$$

Thus

$$X_n = (1-Q_n)[(N\oplus S) + \tilde{A}]Q_n =$$

$$(V_{k_n} \oplus 1_H)(1-P_n)(V_{k_n}^* \oplus 1_H)[(N\oplus S)+\tilde{A}][((1_H-V_{k_n}V_{k_n}^*)\oplus 0) + (V_{k_n}\oplus 1_H)P_n(V_{k_n}^*\oplus 1_H)],$$

and since

$$(V_{k_n}^* \oplus 1_H)[(N\oplus S) + \tilde{A}][(1_H-V_{k_n}V_{k_n}^*) \oplus 0] = 0$$

(recall that $V_{k_n}^* N = NV_{k_n}^*$), we have

$$||X_n|| = ||(V_{k_n}\oplus 1_H)(1-P_n)(V_{k_n}^*\oplus 1_H)[(N\oplus S)+\tilde{A}](V_{k_n}\oplus 1_H)P_n(V_{k_n}^*\oplus 1_H)||$$

$$= ||(1-P_n)(V_{k_n}^*\oplus 1_H)[(N\oplus S) + \tilde{A}](V_{k_n}\oplus 1_H)P_n|| ,$$

since $V_{k_n} \oplus 1_H$ is an isometry and $V_{k_n}^* \oplus 1_H$ is a co-isometry. By doing the indicated matrix multiplications, we obtain

$$||X_n|| = ||(1-P_n)[(N\oplus S) + (V_{k_n}^*\oplus 0)\tilde{A}]P_n||$$

$$\leq ||(1-P_n)(N\oplus S)P_n|| + ||(V_{k_n}^*\oplus 0)\tilde{A}P_n||$$

$$\leq ||(1-P_n)(N\oplus S)P_n|| + 1/n .$$

Thus $||X_n|| \to 0$, proving that $(N\oplus S) + \tilde{A}$ is quasitriangular with implementing sequence of projections $\{Q_n\}$.

Here too it is worth mentioning that a stronger result than Theorem 4.1 is actually valid. Since if N is an arbitrary normal operator, N is unitarily equivalent to a compact perturbation of a diagonable normal operator of uniform infinite multiplicity (cf. the proof of [17], Lemma 1.3), Theorem 4.1 remains valid if N is assumed to be any normal operator whatsoever.

5. The third reduction

It is immediate from Theorems 3.2 and 4.1 that to prove Theorem A, it suffices to prove that every operator acting on $H \oplus H$ of the form $N \oplus S$ is quasitriangular, where N is a diagonable normal operator of uniform multiplicity (and therefore automatically satisfies $E(N) = \sigma(N)$), and where S has the properties that $E(S) = \sigma(N)$ and $\omega(S)$ is empty. We would like to be in the position of knowing that $\sigma(S) \subset \sigma(N)$, and the purpose of this section is to make the needed reduction. The following theorem, which we believe to be new, accomplishes the task. It will be convenient in what follows to define every operator that acts on a finite dimensional space to be quasitriangular, and this we do.

THEOREM 5.1 Let S be any operator on H such that $\omega(S)$ is empty. Then there exist a compact operator K in $L(H)$ and Hilbert spaces K_1 and K_2 with $0 \leq \dim K_1$, $\dim K_2 \leq \aleph_0$, such that $S+K$ is unitarily equivalent to an operator matrix in $L(K_1 \oplus K_2)$ of the form

$$\begin{pmatrix} S_1 & B \\ 0 & S_2 \end{pmatrix},$$

where S_1 is quasitriangular and where S_2 satisfies $\sigma(S_2) \subset E(S)$.

Proof. The set $\mathbb{C}\backslash E(S)$ is an open set which has finitely many (perhaps zero) or countably infinitely many bounded components. Let M be a (perhaps empty) initial segment of the positive integers such that the bounded components of $\mathbb{C}\backslash E(S)$ are exactly the connected open sets $\{U_m\}_{m \in M}$, and (if $M \neq \emptyset$) for each m in M, let λ_m be a fixed point in U_m. Observe that each $S-\lambda_m \in (F)$ and that, by hypothesis, $i(S-\lambda_m) \geq 0$. In case $M \neq \emptyset$, we now construct a compact operator K so that for each m in M, $S+K-\lambda_m$ is right invertible. To do this, we construct a sequence $\{F_m\}_{m \in M}$ of finite rank operators on H as follows. Let F_1 be a finite rank operator such that $||F_1|| < 1$ and such that $S+F_1-\lambda_1$ has range H. Then $S+F_1-\lambda_1$ is right invertible, and since the set of right invertible elements in $L(H)$ is open in the norm topology (cf. [11], p. 35), there exists a number p_1, $0 < p_1 < 1$, with the property that if $X \in L(H)$ and $||X - (S+F_1-\lambda_1)|| < p_1$, then X is also right invertible. Since $i(S+F_1-\lambda_2) = i(S-\lambda_2) \geq 0$, we may now choose a finite rank operator F_2 such that $S+F_1+F_2-\lambda_2$ has range H and such that $||F_2|| < (1/4)p_1$; the latter condition ensures that $S+F_1+F_2-\lambda_1$ is also right invertible. It follows that there exists a number p_2, $0 < p_2 < (1/16)p_1$, such that every operator in the open ball of radius p_2 about $S+F_1+F_2-\lambda_2$ is right invertible, and we continue to construct F_3 in the indicated fashion. Thus we obtain, by induction, a (finite or infinite) sequence $\{F_m\}_{m\in M}$ of finite rank operators on H with the property that the norms of the F_m decrease so rapidly that $\sum_{m\in M} ||F_m|| < +\infty$ and for each k in M, the operator $(S + \sum_{m\in M} F_m) - \lambda_k$ is right invertible. Define $K = \sum_{m\in M} F_m$, observe that K is compact, and write $S' = S+K$. Then $E(S') = E(S)$ and $\omega(S') = \omega(S) = \emptyset$. Thus, by a change of notation, we may simply assume that either $M = \emptyset$ or the operator S of the hypothesis has the property that for each m in M, $S-\lambda_m$ is right invertible. Furthermore, since the intersection of $\sigma(S)$ and the unbounded component B of $\mathbb{C}\backslash E(S)$ is countable, we may also suppose, via a similar argument, that for $\lambda\in\sigma(S)\cap B$, $S-\lambda$ is right invertible.

We now construct by transfinite induction an increasing family $\{M_\alpha\}_{\alpha<\Omega}$ of subspaces of H, indexed by the well ordered set of ordinal numbers α such that $\alpha < \Omega$ (the smallest uncountable ordinal), and having the following properties for each $\alpha < \Omega$:

a) M_α is an invariant subspace for S,

b) $S|M_\alpha$ is quasitriangular, and

c) if $\alpha_1 < \alpha_2 \leq \alpha$, and Q_{α_1,α_2} denotes the projection on the subspace $M_{\alpha_2} \ominus M_{\alpha_1}$, then $Q_{\alpha_1,\alpha_2}SQ_{\alpha_1,\alpha_2}|(M_{\alpha_2}\ominus M_{\alpha_1})$ is quasitriangular.

To begin the construction, we define M_0 to be the subspace of H spanned by all of the eigenvectors of S.

Then M_0 is clearly an invariant subspace for S, and, if $M_0 \neq (0)$, it is not hard to see that there exists an orthonormal basis for M_0, indexed by a subset of the positive integers, such that the matrix for $S|M_0$ relative to this basis is in upper triangular form. Thus $S|M_0$ is quasitriangular. (Alternatively, the fact that $S|M_0$ is quasitriangular is an immediate consequence of Theorem 10.2.) In general, suppose that β is an ordinal number less than Ω, and suppose that we have defined M_α for $\alpha < \beta$ in such a way that a), b) and c) are satisfied for each $\alpha < \beta$. To define M_β, suppose first that β is not a limit ordinal. Then there exists an ordinal number γ such that $\beta = \gamma+1$. We write P for the projection on the subspace $M_\gamma^\perp$, and we define

$$M_\beta = M_\gamma \oplus \vee\{x \in M_\gamma^\perp : PSPx = \lambda_x x \text{ for some } \lambda_x \text{ in } \mathbb{C}\}.$$

To see that $SM_\beta \subset M_\beta$, note that since $SM_\gamma \subset M_\gamma$ by the induction hypothesis, it suffices to show that $S(M_\beta \ominus M_\gamma) \subset M_\beta$. But $M_\beta \ominus M_\gamma$ is spanned by vectors x with the property that $PSPx = \lambda x$, and for such a vector x, $Sx = SPx = PSPx + (1-P)SPx = \lambda x+y$, where $y \in M_\gamma$. Thus M_β is an invariant subspace for S, and a) is satisfied for all ordinal numbers $\alpha \leq \beta$. Next let $Q_{\gamma,\beta}$ be the projection on the subspace $M_\beta \ominus M_\gamma$, and observe that, by definition, $M_\beta \ominus M_\gamma$ is spanned by eigenvectors of the operator $Q_{\gamma,\beta}SQ_{\gamma,\beta}$. Thus, just as before, $Q_{\gamma,\beta}SQ_{\gamma,\beta}|(M_\beta \ominus M_\gamma)$ is quasitriangular. This fact, together with the fact that $S|M_\gamma$ is quasitriangular, implies via [13, Theorem 3] that $S|M_\beta$ is quasitriangular. Thus b) is satisfied for all ordinal numbers $\alpha \leq \beta$. To establish c) for all $\alpha \leq \beta$, it suffices, via the induction hypothesis, to fix an ordinal number $\delta < \gamma$ and to show that $Q_{\delta,\beta}SQ_{\delta,\beta}|(M_\beta \ominus M_\delta)$ is quasitriangular. But, by the induction hypothesis, $Q_{\delta,\gamma}SQ_{\delta,\gamma}|(M_\gamma \ominus M_\delta)$ is quasitriangular, and we have already shown that $Q_{\gamma,\beta}SQ_{\gamma,\beta}|(M_\beta \ominus M_\gamma)$ is quasitriangular. Thus $Q_{\delta,\beta}SQ_{\delta,\beta}|(M_\beta \ominus M_\delta)$ may be regarded as a 2×2 upper triangular operator matrix with quasitriangular diagonal entries, and it follows again from [13, Theorem 3] that $Q_{\delta,\beta}SQ_{\delta,\beta}|M_\beta \ominus M_\delta$ is quasitriangular.

<u>We turn now to the case that</u> β <u>is a limit ordinal</u>. We define $M_\beta = \bigvee_{\alpha<\beta} M_\alpha$, and we observe that M_β is an invariant subspace for S. To verify that $S|M_\beta$ is quasitriangular, it suffices to verify c), i.e., if $\delta < \beta$, then $Q_{\delta,\beta}SQ_{\delta,\beta}|(M_\beta \ominus M_\delta)$ is quasitriangular. For, $S|M_\delta$ is quasitriangular by the induction hypothesis, and the quasitriangularity of $S|M_\beta$ follows once again from Theorem 3 of [13]. To show that $Q_{\delta,\beta}SQ_{\delta,\beta}|(M_\beta \ominus M_\delta)$ is quasitriangular, let $\delta = \alpha_0 < \alpha_1 < \ldots$ be an increasing sequence of ordinal numbers with supremum β. Then, by the induction hypothesis, $Q_{\alpha_i,\alpha_{i+1}}SQ_{\alpha_i,\alpha_{i+1}}|(M_{\alpha_{i+1}} \ominus M_{\alpha_i})$ is quasitriangular for $0 \leq i < \infty$, and since $Q_{\delta,\beta}SQ_{\delta,\beta}|(M_\beta \ominus M_\delta)$ may be regarded as an infinite upper triangular operator matrix with quasitriangular entries on the diagonal, the quasitriangularity of $Q_{\delta,\beta}SQ_{\delta,\beta}|(M_\beta \ominus M_\delta)$ follows from [13, Theorem 4]. Thus we have verified that a), b) and c) are valid for all ordinal numbers $\alpha \leq \beta$.

Hence, by transfinite induction, we obtain an increasing family $\{M_\alpha\}_{\alpha<\Omega}$ of subspaces of H satisfying a), b) and c) for all ordinal numbers $\alpha < \Omega$. Observe now that since H is separable, there mus exist some $\alpha_0 < \Omega$ such that $M_{\alpha_0+1} = M_{\alpha_0}$, and we define the Hilber space K_1 to be the subspace M_{α_0} and the Hilbert space K_2 to be the subspace $H \ominus M_{\alpha_0}$. This identification of H with $K_1 \oplus K_2$ clearly carries S onto an operator matrix acting on $K_1 \oplus K_2$ of the form

$$S' = \begin{pmatrix} S_1 & B \\ 0 & S_2 \end{pmatrix},$$

where S_1 is a quasitriangular operator and S_2 has no point spectrum (for otherwise, $M_{\alpha_0+1} \neq M_{\alpha_0}$). To complete the proof, we show that $\sigma(S_2) \subset E(S)$. Suppose first that $\lambda \notin E(S) (= E(S'))$. Then $(S'-\lambda)$ is Fredholm, and it results easily that there exist operators $X = (X_{ij})$ and $G = (G_{ij})$ on $K_1 \oplus K_2$ with G compact such that $(S'-\lambda)X = 1_{K_1 \oplus K_2} + G$. By doing the appropriate matrix multiplication, we obtain that $(S_2-\lambda)X_{22} = 1_{K_2} + G_{22}$, which implies that $S_2-\lambda$ has closed range and finite dimensional co-kernel. Since $S_2-\lambda$ has no kernel, we conclude that $S_2-\lambda$ is Fredholm, and hence $\lambda \notin E(S_2)$. This shows that $E(S_2) \subset E(S') = E(S)$, and we complete the proof by fixing λ in $\mathbb{C}\backslash E(S')$ and showing that $S_2-\lambda$ must be invertible. We know that $S_2-\lambda$ is Fredholm and that $S_2-\lambda$ has trivial kernel. Thus it suffices to show that the range of $S_2-\lambda$ is all of K_2. If $S'-\lambda$ is invertible, this follows from a straightforward matricial calculation. On the other hand, if $\lambda \in \sigma(S')\backslash E(S') = \sigma(S)\backslash E(S)$, suppose first that λ belongs to one of the bounded components u_k of $\mathbb{C}\backslash E(S)$. Then by the reduction made earlier, we know that there exists a point λ_k in u_k such that $S-\lambda_k$ (and therefore the unitarily equivalent operator $S'-\lambda_k$) has trivial co-kernel. Join λ to λ_k with an arc Γ lying entirely in u_k, observe that $S_2-\lambda_k$ has trivial co-kernel, and conclude that $i(S_2-\lambda_k) = 0$. It follows from the continuity of the index that $i(S_2-\gamma) = 0$ for all $\gamma \in \Gamma$, and, in particular, $i(S_2-\lambda) = 0$. Since $S_2-\lambda$ has trivial kernel, $S_2-\lambda$ is invertible. Similarly, if $\lambda \in \sigma(S) \cap B$ then $S_2-\lambda$ is invertible, so the proof is complete.

As stated in the paragraph preceding the statement of Theorem 5.1, this theorem allows us, once again, to reduce the problem of proving Theorem A.

COROLLARY 5.2 If every operator on $H \oplus H$ of the form $N \oplus S$ is quasitriangular, where N is a diagonable normal operator of uniform infinite multiplicity and $\sigma(S) \subset \sigma(N)$, then Theorem A is true.

Proof. We saw earlier that by virtue of Theorems 3.2 and 4.1, to prove Theorem A it suffices to show that every operator acting on $H \oplus H$ of the form $N \oplus T$ is quasitriangular, where N is as indicated in the statement of the corollary and T satisfies $E(T) = \sigma(N)$ and $\omega(T) = \emptyset$. Thus, let us consider such an operator $N \oplus T$. We apply Theorem 5.1 to T to obtain a compact operator K and Hilbert spaces K_1 and K_2 such that $T+K$ is unitarily equivalent to an operator matrix in $L(K_1 \oplus K_2)$ of the form

$$\begin{pmatrix} S_1 & B \\ 0 & S \end{pmatrix},$$

where S_1 is quasitriangular and where S satisfies $\sigma(S) \subset E(T) = \sigma(N)$. Clearly $N \oplus T$ is quasitriangular if and only if $N \oplus (T+K)$ is also, and $N \oplus (T+K)$ is unitarily equivalent to the operator

$$V = \begin{pmatrix} S_1 & 0 & B \\ 0 & N & 0 \\ 0 & 0 & S \end{pmatrix}$$

acting on the Hilbert space $K_1 \oplus H \oplus K_2$. Moreover, since S_1 is quasitriangular, to show that V is quasitriangular it suffices, by virtue of [13, Theorem 3], to show that $N \oplus S$ is quasitriangular. If K_2 is finite dimensional, then $N \oplus S$ is the sum of the compact operator $0 \oplus S$ and the normal operator $N \oplus 0$, and hence is quasitriangular. Thus we may assume that K_2 has dimension $\aleph_0$, and it follows that K_2 can be identified with H. Hence, the proof is complete.

6. The fourth reduction

By virtue of Corollary 5.2, in order to prove Theorem A it suffices to show that every operator on $H \oplus H$ of the form $N \oplus S$ is quasitriangular, where N is a diagonable normal operator of uniform infinite multiplicity and where S satisfies $\sigma(S) \subset \sigma(N)$. The next important idea of [5] is to replace the (possibly disconnected) compact set $\sigma(N)$ whose complement may have infinitely many bounded components with a slightly larger and more manageable set by making an approximation argument. The topological lemma that is required for this purpose follows, and it is sufficiently standard that we omit its proof.

LEMMA 6.1 Let σ be a nonempty compact subset of the complex plane $\mathbb{C}$, and let ε be any positive number. Then there exists a compact set σ_ε containing σ in its interior such that

1) the boundary of σ_ε consists exactly of the union of $k \geq 1$ disjoint, simple closed, smooth Jordan curves $\Gamma_1,\ldots,\Gamma_k$, and
2) if ζ_1 is any point in σ_ε, then there exists a point ζ_2 in σ such that $|\zeta_1-\zeta_2| < \varepsilon$.

As the reader may observe, there are two important properties that the compact sets $\sigma_\varepsilon(\varepsilon>0)$ have. The first is that σ is contained in the interior of each σ_ε, and the second is that σ_ε has only finitely many (connected) components, each of which has a smooth boundary. The utility of the above lemma becomes quickly apparent.

PROPOSITION 6.2. Let N be a diagonable normal operator on H of uniform infinite multiplicity with spectrum $\sigma = \sigma(N)$, and let $\{\varepsilon_n\}$ be a monotone sequence of positive numbers converging to 0. Then for each positive integer n, there exists a diagonable normal operator N_n on H of uniform infinite multiplicity such that $\sigma(N_n)$ is the set σ_{ε_n} defined in Lemma 6.1 and such that $||N_n-N|| < \varepsilon_n$. Furthermore, if S is an operator on H such that $N_n \oplus S$ is quasitriangular for each positive integer n, then $N \oplus S$ is quasitriangular also.

Proof. It was established in [14] that the set of quasitriangular operators is closed in the uniform operator topology, so the last assertion of the proposition follows from the first. Thus, let n be a fixed positive integer, and let σ_{ε_n} be the compact set containing σ provided by Lemma 6.1. Let $\{\lambda_m\}_{m\in M}$ be a sequence consisting of the distinct eigenvalues of N (perhaps σ is a single point), and let $\{\beta_k\}_{k=1}^{\infty}$ be a countable dense set in σ_{ε_n}. For each fixed positive integer m in M, let $\{\beta_{m,j}\}_{j\in J_m}$ be the (infinite) subset of the set $\{\beta_k\}$ such that $|\beta_{m,j}-\lambda_m| < \varepsilon_n$, $j \in J_m$. (Since the set $\{\beta_k\}$ is dense in σ_{ε_n}, and the interior of σ_{ε_n} contains σ, there must be infinitely many such $\beta_{m,j}$.) It is now a triviality to construct a diagonable normal operator D_m of uniform infinite multiplicity, acting on the subspace $\mathcal{M}_m$ of $\mathcal{H}$ on which N is the scalar operator λ_m, and having the property that the eigenvalues of D_m are exactly the numbers $\{\beta_{m,j}\}_{j\in J_m}$. We define $N_n = \sum_{m\in M} \oplus D_m$, and we observe that clearly N_n is diagonable and of uniform multiplicity; by construction, $||N-N_n|| < \varepsilon_n$. What is left to verify is that $\sigma(N_n) = \sigma_{\varepsilon_n}$. This is a consequence of the facts that the sequence $\{\beta_k\}_{k=1}^{\infty}$ is dense in σ_{ε_n} and each β_k must lie within ε_n of some λ_m, since by 2) of Lemma 6.1, $\text{dist}(\beta_k,\sigma) < \varepsilon_n$ and the sequence $\{\lambda_m\}_{m\in M}$ is dense in σ.

7. The fifth reduction

It follows from the results of Section 3-6 that to prove Theorem A, it suffices to prove that every operator on $\mathcal{H} \oplus \mathcal{H}$ of the form $N \oplus S$ is quasitriangular, where N is a diagonable normal operator of uniform infinite multiplicity such that $\partial\sigma(N)$ consists of a finite number of disjoint, simple, closed, smooth, Jordan curves, and where S is an operator on $\mathcal{H}$ whose spectrum is contained in the interior of $\sigma(N)$. If N and S are such operators, then, of course, it is possible that $\sigma(N) (= \sigma(N\oplus S))$ is not connected, but rather is the disjoint union of finitely many compact connected subsets. The purpose of the present section is to carry out the reduction that will allow us to assume that $\sigma(N)$ is connected. The following lemma is well known; we include its proof simply for completeness.

LEMMA 7.1. Let S be an operator on H whose spectrum consists of the union of m disjoint, nonempty, compact sets $\kappa_1,\ldots,\kappa_m$. Then S is similar to an orthogonal direct sum $S_1 \oplus \ldots \oplus S_m$, where for $1 \leq i \leq m$, $\sigma(S_i) = \kappa_i$.

Proof. If $m = 1$, there is nothing to prove. If $m > 1$, then the compact set κ_1 is a positive distance from the compact set $\kappa_2 \cup \ldots \cup \kappa_m$, and by employing the Riesz functional calculus, one obtains easily an idempotent E in $L(H)$ that commutes with S and has the property that $\sigma(S|\text{range } E) = \kappa_1$. It is elementary that there exists an invertible operator W on H such that WEW^{-1} is a projection with range equal to the range of E, and it follows easily that WSW^{-1} is an orthogonal direct sum $S_1 \oplus S'$ where $\sigma(S_1) = \kappa_1$ and $\sigma(S') = \kappa_2 \cup \ldots \cup \kappa_m$. The result now follows by induction.

Henceforth, the interior of a plane set σ will be denoted by σ^0.

THEOREM 7.2. Suppose that N is a normal operator on H such that $\sigma(N) = \sigma_1 \cup \ldots \cup \sigma_m$, where the σ_i are disjoint, nonempty, compact sets, and suppose that $S \in L(H)$ and satisfies $\sigma(S) \subset \sigma(N)^0$. Then $N \oplus S$ is similar to a direct sum $(N_1 \oplus S_1) \oplus \ldots \oplus (N_m \oplus S_m)$ where for $1 \leq i \leq m$, N_i is a normal operator satisfying $\sigma(N_i) = \sigma_i$ and either the direct summand S_i is absent or $\sigma(S_i) \subset \sigma(N_i)^0$. Furthermore, if each $N_i \oplus S_i$ is quasitriangular, then so is $N \oplus S$.

Proof. The last assertion of the theorem follows immediately from the first assertion and Theorems 3 and 9 of [13]. To prove the first assertion, we apply Lemma 7.1 separately to N and S, observe that the spectral idempotents associated with a normal operator are already projections, and reorder the appropriate direct sums.

8. The sixth reduction

It follows from the results of Sections 3-7 that to prove Theorem A it suffices to show that every operator on $\mathcal{H} \oplus \mathcal{H}$ of the form $N \oplus S$ is quasitriangular, where

1) N is a diagonable normal operator of uniform infinite multiplicity such that $\sigma(N)$ is connected and such that $\partial\sigma(N)$ consists of the disjoint union of a finite number of simple closed, smooth Jordan curves, and

2) $\sigma(S) \subset \sigma(N)^0$.

The next construction is one that allows us to assume that the operator S in the direct sum $N \oplus S$ has a certain sort of "cyclic" vector. If Ω is an open plane set, we shall henceforth denote by $\mathrm{Rat}(\Omega)$ the set of all rational functions $r(z)$ whose poles lie outside Ω. A vector x in a Hilbert space $\mathcal{K}$ is said to be a <u>$\mathrm{Rat}(\Omega)$-cyclic</u> vector for an operator T in $\mathcal{L}(\mathcal{K})$ with $\sigma(T) \subset \Omega$ if the set $\{r(T)x : r \in \mathrm{Rat}(\Omega)\}$ is dense in $\mathcal{K}$.

THEOREM 8.1. Let T be in $\mathcal{L}(\mathcal{H})$ and let Ω be an open plane set with $\sigma(T) \subset \Omega = \Omega^{-0}$. Then there exists a (finite or infinite) sequence M of positive integers and an orthogonal sequence $\{\mathcal{H}_i\}_{i\in M}$ of (finite or infinite dimensional) subspaces $\mathcal{H}_i$ of $\mathcal{H}$ such that

a) $\sum_{i\in M} \oplus \mathcal{H}_i = \mathcal{H}$,

b) The unique matrix $(T_{ij})_{i,j\in M}$ with operator entries $T_{i,j} : \mathcal{H}_j \to \mathcal{H}_i$ that represents T relative to the decomposition a) is in upper triangular form (i.e., $T_{ij} = 0$ for $i > j$),

c) For each i in M, the diagonal entry T_{ii} acting on the Hilbert space $\mathcal{H}_i$ possesses a $\mathrm{Rat}(\Omega)$-cyclic vector, and

d) For each i in M, $\sigma(T_{ii}) \subset \Omega$.

Proof. This theorem is very like [13, Theorem 5], so we only make some remarks about how the proof of that theorem can be altered to prove the present theorem. In the notation of [13, Theorem 5], the space H_1 is now defined to be $\vee\{r(T)x_1 : r \in \mathrm{Rat}(\Omega)\}$. This implies that if $\lambda \notin \overline{\Omega}$, then $(T-\lambda)H_1 \subset H_1$ and $(T-\lambda)^{-1}H_1 \subset H_1$, so $\lambda \notin \sigma(T|H_1)$. If P is the projection on the subspace $H_1^{\perp}$, then it follows that $\sigma(T|H_1) \subset \overline{\Omega}$ and that $\sigma(PTP \mid H_1) \subset \overline{\Omega}$. Since the spectrum of these operators can be enlarged only by bounded components of $\mathbb{C}\backslash\sigma(T)$ and since $\overline{\Omega}^0 = \Omega$, it follows that $\sigma(T|H_1) \subset \Omega$ and $\sigma(PTP|H_1^{\perp}) \subset \Omega$. The induction now proceeds along the lines indicated in [13], with only the obvious changes to be made.

THEOREM 8.2. Let N be a diagonable normal operator on H of uniform infinite multiplicity, let S be any operator on H, and let Ω be an open plane set with $\sigma(S) \subset \Omega = \overline{\Omega}^0$. Then for some cardinal number n satisfying $1 \leq n \leq \aleph_0$, $N \oplus S$ is unitarily equivalent to an $n \times n$ operator matrix (D_{ij}) that is in upper triangular form and has the property that each diagonal entry D_{ii} is of the form $N \oplus S_{ii}$, where S_{ii} has a $\mathrm{Rat}(\Omega)$-cyclic vector and satisfies $\sigma(S_{ii}) \subset \Omega$. Furthermore, if every diagonal entry $N \oplus S_{ii}$ of (D_{ij}) is quasitriangular, then so is $N \oplus S$.

Proof. The last assertion of the theorem is an immediate consequence of [13, Theorems 3 and 4]. To see that $N \oplus S$ is unitarily equivalent to an operator matrix (D_{ij}) of the indicated form, we apply Theorem 8.1 to S to write it as an $n \times n$ upper triangular operator matrix (S_{ij}) where $1 \leq n \leq \aleph_0$, and where each diagonal entry S_{ii} has a $\mathrm{Rat}(\Omega)$-cyclic vector and satisfies $\sigma(S_{ii}) \subset \Omega$. Next, observe that the hypotheses on N guarantee that N is unitarily equivalent to a direct sum of n copies of itself. The result now follows by "judiciously blending" the matrix (S_{ij}) with the $n \times n$ diagonal matrix each of whose diagonal entries is equal to N . We give an illustrative example; namely, the case $n = 2$. The matrix

$$\begin{pmatrix} N & 0 & 0 & 0 \\ 0 & N & 0 & 0 \\ 0 & 0 & S_{11} & S_{12} \\ 0 & 0 & 0 & S_{22} \end{pmatrix}$$

is unitarily equivalent to the matrix

$$\begin{pmatrix} N & 0 & 0 & 0 \\ 0 & S_{11} & 0 & S_{12} \\ 0 & 0 & N & 0 \\ 0 & 0 & 0 & S_{22} \end{pmatrix}$$

We close this section by remarking that if N is a normal operator and S_{ii} is an operator acting on a finite dimensional space, then $N \oplus S_{ii}$ is a compact perturbation of the quasitriangular operator $N \oplus 0$, and therefore is quasitriangular. Thus, to apply Theorem 8.2 to prove that an operator $N \oplus S$ is quasitriangular, it suffices to show that each of the resulting diagonal entries $N \oplus S_{ii}$ is quasitriangular, where S_{ii} acts on a Hilbert space of dimension $\aleph_0$.

9. The seventh reduction

If Γ is a simple closed Jordan curve in $\mathbb{C}$, we denote the open set that is the bounded [resp., unbounded] component of $\mathbb{C}\setminus\Gamma$ by $\mathrm{Int}(\Gamma)$ [resp., $\mathrm{Ext}(\Gamma)$]. We now fix $k \geq 1$ disjoint, simple closed, smooth Jordan curves $\Gamma_1,\ldots,\Gamma_k$ in $\mathbb{C}$ with the properties that

a) if $k > 1$, then $\Gamma_2 \cup\ldots\cup \Gamma_k \subset \mathrm{Int}(\Gamma_1)$, and

b) if $1 < i,j \leq k$ and $i \neq j$, then $\Gamma_i \subset \mathrm{Ext}(\Gamma_j)$.

Let Ω denote the open connected set $\mathrm{Int}(\Gamma_1) \cap \mathrm{Ext}(\Gamma_2) \cap\ldots\cap \mathrm{Ext}(\Gamma_k)$, and observe that $\partial\Omega = \Gamma_1 \cup\ldots\cup \Gamma_k$ and thus that $\overline{\Omega}$ is the disjoint union $\Omega \cup \partial\Omega$. It follows from the results of Sections 3-8 that to prove Theorem A, it suffices to show that every operator on $H \oplus H$ of the form $N \oplus S$ is quasitriangular, where

1) N is a diagonable normal operator of uniform infinite multiplicity satisfying $\sigma(N) = \overline{\Omega}$, and

2) S has a $\mathrm{Rat}(\Omega)$-cyclic vector x_o and satisfies $\sigma(S) \subset \sigma(N)^0 = \Omega$.

The next important idea due to the authors of [5] is that of replacing the operator S in the direct sum $N \oplus S$ by an operator about which more is known. To accomplish this, it will be necessary to introduce some Hilbert spaces that are closely related to the set $\overline{\Omega} = \Omega \cup \partial\Omega$. (Keep in mind the possibility that $k = 1$, i.e., that there is only one curve Γ_1. In this case, of course, $\sigma(N) = \Gamma_1 \cup \mathrm{Int}(\Gamma_1)$, and thus $\sigma(N)$ is simply connected.)

DEFINITION. We denote by $L_2(\partial\Omega)$ the Hilbert space consisting of all those complex functions f defined on $\partial\Omega$ that are square integrable with respect to arc length measure μ on $\partial\Omega$. We also denote by $H_2(\partial\Omega)$ the subspace of $L_2(\partial\Omega)$ consisting of the closure in $L_2(\partial\Omega)$ of the linear manifold $\{r|\partial\Omega : r \in \mathrm{Rat}(\overline{\Omega})\}$. Furthermore, if g is any essentially bounded, measurable function defined on $\partial\Omega$, we denote by M_g the (normal) operator on $L_2(\partial\Omega)$ defined by $M_g f = gf$, and in particular, we denote the multiplication operator $f(z) \to zf(z)$ on $L_2(\partial\Omega)$ by M_z.

Frequently, to avoid cumbersome notational problems, we shall write elements of $H_2(\partial\Omega) \cap \mathrm{Rat}(\overline{\Omega})$ as r instead of $r|\partial\Omega$. The following proposition contains some known facts about these Hilbert spaces that we shall need.

PROPOSITION 9.1. The following statements are valid:

I) $H_2 = H_2(\partial\Omega)$ is a proper subspace of $L_2 = L_2(\partial\Omega)$.

II) M_z is a normal operator on L_2 satisfying $\sigma(M_z) = E(M_z) = \partial\Omega$.

III) The subspace H_2 is invariant for M_z, and if we write M_z^+

for the operator $M_z|H_2$, then M_z^+ is a subnormal operator satisfying $\sigma(M_z^+) = \overline{\Omega}$.

IV) For each λ_o in Ω , $(z-\lambda_o)H_2$ is a closed subspace of H_2 , and $\dim(H_2 \ominus (z-\lambda_o)H_2) = 1$; thus, $E(M_z^+) = \partial\Omega$.

V) The self-commutator $(M_z^+)(M_z^+)^* - (M_z^+)^*(M_z^+)$ is compact.

VI) For each λ_o in Ω , there exists a bounded linear functional E_{λ_o} on H_2 with the property that if $r \in Rat(\overline{\Omega})$, then $E_{\lambda_o}(r) = r(\lambda_o)$. Furthermore, the mapping $\lambda_o \to ||E_{\lambda_o}||$ is a continuous function on Ω .

Proof. That H_2 is a proper subspace of L_2 is well-known and may be verified by consulting [18, page 371]. (Actually, to prove Theorem A, it is not strictly necessary to know this. One could proceed by considering the (in reality, vacuous) case that $H_2 = L_2$.) That II) is correct is completely obvious, as is the fact that H_2 is an invariant subspace for M_z . Since M_z^+ is subnormal and $(z-\lambda)^{-1}H_2 \subset H_2$ for $\lambda \notin \overline{\Omega}$, it follows that $\sigma(M_z^+)$ must be either $\partial\Omega$ or $\overline{\Omega}$. If $\sigma(M_z^+) = \partial\Omega$, then for each λ in Ω , $M_{z-\lambda}(H_2) = H_2$, and thus $(z-\lambda)^{-1} \in H_2$. Hence,

$$\{r|\partial\Omega : r \in Rat(\partial\Omega)\} \subset H_2 ,$$

and since one knows from the Hartogs-Rosenthal theorem that $Rat(\partial\Omega)$ is dense in the space $C(\partial\Omega)$ of continuous complex functions on $\partial\Omega$, it follows that $H_2 = L_2$, which contradicts I). Thus $\sigma(M_z^+) = \overline{\Omega}$.

To prove IV), observe that if $\lambda_o \in \Omega$, then $M_z - \lambda_o = M_{z-\lambda_o}$ is invertible and hence bounded below. Thus, $M_{z-\lambda_o}(H_2) = (M_z^+ - \lambda_o)H_2 = (z-\lambda_o)H_2$ is a closed subspace of H_2 . Since $\lambda_o \in \sigma(M_z^+)$ from III), it cannot be the case that $(z-\lambda_o)H_2 = H_2$, and to show that $H_2 \ominus (z-\lambda_o)H_2$ is one-dimensional, it suffices to show that the sum of the one-dimensional subspace $\{\lambda\}$ of constant functions in H_2 and $(z-\lambda_o)H_2$ is all of H_2 . To accomplish this, let $r \in Rat(\overline{\Omega})$

and let $z \in \partial\Omega$. Then

$$r(z) = (z-\lambda_o)[(r(z) - r(\lambda_o))(z-\lambda_o)^{-1}] + r(\lambda_o) ,$$

and clearly $h(z) = (r(z) - r(\lambda_o))(z-\lambda_o)^{-1} \in H_2$. Thus $\{\lambda\} + (z-\lambda_o)H_2$, which must be closed, is dense in H_2 and hence is all of H_2 . This proves that $H_2 \ominus (M_z^+ - \lambda_o)H_2$ is one-dimensional, and it follows immediately that $M_z^+ - \lambda_o$ is a Fredholm operator of index -1. Thus, $E(M_z^+) = \partial\Omega$.

To prove V), one may simply quote [6], since M_z^+ has the analytic cyclic vector 1, or one may proceed directly as follows. Note that $\pi(M_z^+)$ is a hyponormal element of the Calkin algebra with spectrum $E(M_z^+) = \partial\Omega$, and recall that hyponormal operators with thin spectra are known to be normal. Thus $\pi(M_z^+)$ is normal, which implies that the self-commutator of M_z^+ is compact.

Finally, to prove VI), let λ_o be fixed in Ω . It follows from the proof of IV) that $\dim(H_2 \ominus (z-\lambda_o)H_2) = 1$ and that the identity function $1 = 1|\partial\Omega$ in H_2 does not belong to $(z-\lambda_o)H_2$. Thus, we may write $1 = f_{\lambda_o} + g_{\lambda_o}$ where f_{λ_o} is a non-zero vector in $H_2 \ominus (z-\lambda_o)H_2$ and where $g_{\lambda_o} \in (z-\lambda_o)H_2$. If $r \in \mathrm{Rat}(\overline{\Omega})$, then as in the proof of IV), we write $r(z) = (z-\lambda_o)h(z) + r(\lambda_o)$ where $h \in H_2$, and compute:

$$(r,f_{\lambda_o}) = (r(\lambda_o),f_{\lambda_o}) = r(\lambda_o)(1,f_{\lambda_o}) = r(\lambda_o)||f_{\lambda_o}||^2 .$$

Thus, if we define the linear functional E_{λ_o} on H_2 by $E_{\lambda_o}(k) = (k,f_{\lambda_o}/||f_{\lambda_o}||^2)$, $k \in H_2$, then clearly E_{λ_o} is a bounded linear functional on H_2 , and if $r \in \mathrm{Rat}(\overline{\Omega})$, then $E_{\lambda_o}(r) = r(\lambda_o)$. Furthermore, $||E_{\lambda_o}|| = ||f_{\lambda_o}|| = ||1-g_{\lambda_o}|| = \inf_{k\in H_2}||1-(z-\lambda_o)k||$, from which it follows immediately that the mapping $\lambda_o \to ||E_{\lambda_o}||$ is continuous on Ω . Thus, the proof of Proposition 9.1 is complete.

We introduce now some additional notation.

Consider the decomposition $L_2 = H_2 \oplus (L_2 \ominus H_2)$, and recall from I) above that the subspace $L_2 \ominus H_2$ is nontrivial. We can thus write the operator M_z on L_2 as a 2×2 matrix relative to this decomposition of L_2 - say

$$M_z = \begin{pmatrix} M_z^+ & G \\ 0 & M_z^- \end{pmatrix} .$$

Note that, by definition, M_z^- is the operator $PM_zP|(L_2 \ominus H_2)$, where P is the projection whose range is $L_2 \ominus H_2$. Observe also that since M_z is normal, we have $[M_z^+, (M_z^+)^*] = -GG^*$, and since M_z^+ has a compact self-commutator by V) above, G must be compact. These observations lead to the following proposition.

PROPOSITION 9.2. Let N and S be the operators on $\mathcal{H}$ as described in 1) and 2) at the beginning of this section, and let M_z, M_z^+, and M_z^- be as defined above. Then the operator $N \oplus S$ is unitarily equivalent to a compact perturbation of the operator $(N \oplus M_z^+) \oplus (M_z^- \oplus S)$, and thus if both $N \oplus M_z^+$ and $M_z^- \oplus S$ are quasitriangular, then so is $N \oplus S$.

Proof. The last assertion of the proposition is an immediate consequence of [14] and the first assertion, so it suffices to prove the first assertion. We begin by observing that the operator $N \oplus M_z$ acting on the Hilbert space $\mathcal{H} \oplus L_2(\partial\Omega)$ is normal and satisfies $E(N \oplus M_z) = E(N) \cup E(M_z) = \overline{\Omega} = E(N)$. It follows from the converse to the Berg-von Neumann-Weyl theorem that N is unitarily equivalent to a compact perturbation of $N \oplus M_z$. Furthermore, we saw above that M_z is a compact perturbation of the direct sum $M_z^+ \oplus M_z^-$ acting on the Hilbert space $H_2 \oplus (L_2 \ominus H_2)$. Thus, if we write $A \sim B$ to mean that A is unitarily equivalent to a compact perturbation of B, then we have

$$N \oplus S \sim (N \oplus M_z) \oplus S \sim N \oplus (M_z^+ \oplus M_z^-) \oplus S \sim (N \oplus M_z^+) \oplus (M_z^- \oplus S) ,$$

and the proof is complete.

It follows from the foregoing that to prove Theorem A, it suffices to show that each of the operators $N \oplus M_z^+$ and $M_z^- \oplus S$ is quasitriangular. Different techniques are needed to handle these two operators, and we attack the operator $M_z^- \oplus S$ first. The next theorem contains another essential idea from [5]; it allows us to replace the operator S in the direct sum $M_z^- \oplus S$ with a more manageable operator. We continue to use the notation and terminology introduced earlier in this section.

THEOREM 9.3. Let S be an operator in $L(H)$ satisfying $\sigma(S) \subset \Omega$ and possessing a $\mathrm{Rat}(\Omega)$-cyclic vector x_0. Then there exists a bounded linear transformation $Y : H_2(\partial\Omega) \to H$ whose range is dense in H such that $YM_z^+ = SY$.

Proof. For r in $\mathrm{Rat}(\overline{\Omega})$, we define $Y(r|\partial\Omega) = r(S)x_o$. (Note that since $\sigma(S) \subset \Omega$, $r(S)$ is a well-defined rational function of S, and clearly $r|\partial\Omega$ determines r.) Thus Y is clearly a linear transformation of $H_2 \cap \mathrm{Rat}(\overline{\Omega})$ into H whose range is dense in H that satisfies

$$(YM_z^+)r(z) = Y(zr(z)) = Sr(S)x_o = (SY)r(z) , \quad r \in \mathrm{Rat}(\overline{\Omega}) ,$$

and thus to complete the proof it suffices to show that Y is bounded on $H_2 \cap \mathrm{Rat}(\overline{\Omega})$. We do this by showing that there exists a constant K that satisfies $||Yr||_H \le K||r||_{H_2}$ for each r in $\mathrm{Rat}(\overline{\Omega})$. The argument goes as follows. Clearly

$$||Yr||_H = ||r(S)x_o||_H \le ||r(S)|| \; ||x_o|| ,$$

so that it suffices to show that there exists a constant K_2 such that $||r(S)|| \le K_2||r||_{H_2}$ for all r in $\mathrm{Rat}(\overline{\Omega})$. Since $\sigma(S)$ is contained in the open set Ω, there exist a finite number of (propertly oriented) simple closed Jordan curves $\gamma_1,\ldots,\gamma_k$ contained in $\Omega\setminus\sigma(S)$ such that

$$r(S) = \frac{1}{2\pi i}\int_{\gamma_1\cup\ldots\cup\gamma_k} r(\zeta)(\zeta-S)^{-1}d\zeta \ , \ r \in \mathrm{Rat}(\overline{\Omega}) \ .$$

It follows easily from this representation that there exists a constant K_1 (independent of r) such that

$$||r(S)|| \leq K_1 \max\{|r(\zeta)| : \zeta \in \gamma_1 \cup\ldots\cup \gamma_k\} \ , \ r \in \mathrm{Rat}(\overline{\Omega}) \ .$$

Using VI) of Proposition 9.1, we obtain

$$\begin{aligned} ||r(S)|| &\leq K_1 \max\{|E_\zeta(r)| : \zeta \in \gamma_1 \cup\ldots\cup \gamma_k\} \\ &\leq K_1 \max\{||E_\zeta||\ ||r||_{H_2} : \zeta \in \gamma_1 \cup\ldots\cup \gamma_k\} \\ &= K_1(\max\{||E_\zeta|| : \zeta \in \gamma_1 \cup\ldots\cup \gamma_k\})||r||_{H_2} \\ &\leq K_2||r||_{H_2} \ , \end{aligned}$$

since the map $\zeta \to ||E_\zeta||$ is continuous on Ω, and $\gamma_1\cup\ldots\cup\gamma_k$ is a compact subset of Ω. This completes the proof of Theorem 9.3. (The last part of this argument is similar to an argument in [6]).

Theorem 9.3 is the key, together with the material of Section 10, that allows us to prove that $S \oplus M_z^-$ is quasitriangular.

COROLLARY 9.4. Let S, Y and M_z^+ be as in Theorem 9.3 and let M_z^- and M_z be as defined earlier. Then there exists a compact operator K in $\mathcal{L}(\mathcal{H} \oplus (L_2 \ominus H_2))$ and an operator $\tilde{Y} : L_2 \to \mathcal{H} \oplus (L_2 \ominus H_2)$ whose range is dense in $\mathcal{H} \oplus (L_2 \ominus H_2)$ such that $\tilde{Y}M_z = ((S \oplus M_z^-) + K)\tilde{Y}$.

Proof. We have from Theorem 9.3 that $YM_z^+ = SY$, and we recall from earlier that the normal operator M_z on $L_2(\partial\Omega)$ can be written as the 2×2 operator matrix

$$M_z = \begin{pmatrix} M_z^+ & G \\ 0 & M_z^- \end{pmatrix}$$

where $G : (L_2 \ominus H_2) \to H_2$ is a compact operator. Define K to be the compact operator on $\mathcal{H} \oplus (L_2 \ominus H_2)$ given by the matrix

$$\begin{pmatrix} 0 & YG \\ 0 & 0 \end{pmatrix},$$

and define $\tilde{Y} : L_2 \to \mathcal{H} \oplus (L_2 \ominus H_2)$ by $\tilde{Y} = Y \oplus 1_{(L_2 \ominus H_2)}$. It is obvious that Y has range dense in $\mathcal{H} \oplus (L_2 \ominus H_2)$, and that $\tilde{Y}M_z = ((S \oplus M_z^-) + K)Y$ follows from the following elementary calculation:

$$\begin{pmatrix} Y & 0 \\ 0 & 1 \end{pmatrix} \begin{pmatrix} M_z^+ & G \\ 0 & M_z^- \end{pmatrix} = \begin{pmatrix} S & YG \\ 0 & M_z^- \end{pmatrix} \begin{pmatrix} Y & 0 \\ 0 & 1 \end{pmatrix} .$$

10. The proof that $S \oplus M_z^-$ is quasitriangular

We saw in Section 9 that to prove Theorem A, it suffices to prove (in the notation of that section) that each of the operators $N \oplus M_z^+$ and $S \oplus M_z^-$ is quasitriangular. To prove that $S \oplus M_z^-$ is quasitriangular requires another deep insight due to the authors of [5]. An operator T on a separable Hilbert space $\mathcal{K}$ is said to have the <u>spectral splitting property</u> if for every $\varepsilon > 0$,

$$\bigvee\{x \in \mathcal{K} : \inf_{\lambda \in \mathbb{C}} \ \underline{\lim}_{n \to \infty} ||(T-\lambda)^n x||^{1/n} < \varepsilon\} = \mathcal{K}$$

(where as usual, $\bigvee \mathcal{S}$ denotes the closed linear manifold generated by the set $\mathcal{S}$). It is an easy consequence of the spectral theorem that every normal operator (on a separable space) has the spectral splitting property. Moreover, the following result concerning preservation of the spectral splitting property is sufficiently obvious that we omit its proof.

PROPOSITION 10.1. Suppose that K_1 and K_2 are separable Hilbert spaces, and let A and B be operators on K_1 and K_2, respectively such that A has the spectral splitting property. If there exists an operator $\tilde{Y} : K_1 \to K_2$ whose range is dense in K_2 and satisfies $\tilde{Y}A = B\tilde{Y}$, then B also has the spectral splitting property.

Since the operator M_z acting on the Hilbert space $L_2 = L_2(\partial\Omega)$ is normal, it follows immediately from Corollary 9.4 and Proposition 10.1 that the operator $(S \oplus M_z^-) + K$ (where K is as defined in Corollary 9.4) has the spectral splitting property. Since K is a compact operator, the proof that $S \oplus M_z^-$ is quasitriangular is completed by an application of the following theorem, which is due to Apostol [1].

THEOREM 10.2. If T is an operator on H that has the spectral splitting property, then T is quasitriangular.

Proof. Let $\varepsilon > 0$ be fixed. Since the set of quasitriangular operators in $L(H)$ is closed [14], to prove the theorem it suffices to construct a quasitriangular operator T_ε such that $||T - T_\varepsilon|| < \varepsilon$ We consider first the case that for a given λ in $\mathbb{C}$, $T-\lambda$ possesses a cyclic vector x_o with the property that $\underline{\lim}_n ||(T-\lambda)^n x_o||^{1/n} < \varepsilon$, and it is easy to see that there is no generality lost by assuming that $||x_o|| = 1$. Consider the sequence of vectors $\{x_o, (T-\lambda)x_o, (T-\lambda)^2 x_o, \ldots\}$ whose span is H, and let $\{e_n\}_{n=0}^{\infty}$ be the orthonormal basis for H obtained by applying the Gram-Schmidt procedure to this sequence. Then the matrix $(\alpha_{ij})_{i,j=0}^{\infty}$ for $T-\lambda$ with respect to the orthonormal basis $\{e_n\}$ clearly has the property that it is almost upper triangular, in the sense that the only diagonal below the main diagonal that contains nonzero entries is the diagonal immediately below the main diagonal. Furthermore (since $x_o = e_o$), an easy matricial calculation shows that for every positive integer n ,

$$((T-\lambda)^n x_o, e_n) = ((T-\lambda)^n e_o, e_n) = \alpha_{1,0}\alpha_{2,1}, \ldots, \alpha_{n,n-1} \quad ,$$

and thus the hypothesis guarantees that

$$\varliminf_n |\alpha_{1,0}\alpha_{2,1}, \ldots, \alpha_{n,n-1}|^{1/n} < \varepsilon \ .$$

Since x_o is a cyclic vector for $T-\lambda$, no $\alpha_{i,i-1}$ can be zero, and it follows readily that there exists a subsequence $\{\alpha_{n_k}, \alpha_{n_k-1}\}_{k=1}^{\infty}$ of entries on this diagonal each having modulus less than ε . Thus if we define the operator $T_\varepsilon - \lambda$ to be the matrix obtained from the matrix of $T-\lambda$ by replacing all of the entries α_{n_k, n_k-1} by 0 , then T_ε is clearly quasitriangular and satisfies $||T-T_\varepsilon|| < \varepsilon$.

We turn now to the general case. Since T has the spectral splitting property, there exist countable sets $\{x_n\}_{n=1}^{\infty}$ in H and $\{\lambda_n\}_{n=1}^{\infty}$ in $\mathbb{C}$ such that $\bigvee\{x_n\} = H$ and such that for each integer k , $1 \leq k < \infty$, $\varliminf_n ||(T-\lambda_k)^n x_k||^{1/n} < \varepsilon$. We shall construct a (finite or infinite) sequence $\{M_j\}_{j\in J}$ consisting of mutually orthogonal subspaces of H indexed by an initial segment J of the positive integers and having the following properties:

α) $\sum_{j\in J} \oplus M_j = H$,

β) for each j in J , $\sum_{i\leq j} \oplus M_i$ is an invariant subspace for T , and

γ) for each j in J , there exists a quasitriangular operator Q_j in $L(M_j)$ such that $||(P_j T P_j | M_j) - Q_j|| < \varepsilon$, where P_j is the projection in $L(H)$ with range M_j .

Suppose, for the moment, that the sequence $\{M_j\}_{j\in J}$ has already been constructed. Then the operator T can be regarded as a matrix with operator entries $(T_{ij})_{i,j\in J}$ relative to the decomposition $H = \sum_{j\in J} \oplus M_j$, and it follows from β) that the matrix (T_{ij}) is in upper triangular form. Furthermore, it follows from γ) that for every j in J , $||T_{jj} - Q_j|| < \varepsilon$. Thus, the operator T_ε obtained by replacing each diagonal entry T_{jj} in the matrix (T_{ij}) by the corresponding operator Q_j is quasitriangular [13, Theorems 3 and 4] and satisfies $||T - T_\varepsilon|| < \varepsilon$. Thus, the proof can be completed by constructing the sequence $\{M_j\}_{j\in J}$ satisfying α), β) and γ).

Let $M_1 = \bigvee\{x_1, (T-\lambda_1)x_1, (T-\lambda_1)^2 x_1, \ldots\}$, and observe that M_1 is an invariant subspace for T . Furthermore, by what was proved first (or by definition if M_1 is finite dimensional), it is clear that there exists a quasitriangular operator Q_1 in $L(M_1)$ such that $||Q_1 - (T|M_1)|| < \varepsilon$. If $M_1 = H$, the construction is complete, so suppose $M_1 \neq H$, and let x_k be the first vector in the sequence $\{x_n\}_{n=1}^{\infty}$ that is not contained in M_1 . We know that $\underline{\lim}_n ||(T-\lambda_k)^n x_k||^{1/n} < \varepsilon$, and for our purposes there is no loss of generality in assuming that $\lambda_k = 0$. Let P denote the projection on the subspace $M_1^\perp$, note that $Px_k \neq 0$, and consider the following equation, whose validity depends upon the fact that $PT(1-P) = 0$:

$$(PTP)^n (Px_k) = PT^n Px_k = PT^n x_k - PT^n (1-P) x_k = PT^n x_k \ .$$

Thus $\underline{\lim}_n ||(PTP)^n (Px_k)||^{1/n} \leq \underline{\lim}_n ||T^n x_k||^{1/n} < \varepsilon$, and we proceed to define M_2 as

$$M_2 = \bigvee\{Px_k, (PTP)(Px_k), (PTP)^2 (Px_k), \ldots\} \ .$$

That $M_1 \oplus M_2$ is an invariant subspace for T follows by an argument like one employed in Theorem 5.1, and the proof that there is a quasitriangular operator Q_2 in $L(M_2)$ such that $||(P_2 T P_2 | M_2) - Q_2|| < \varepsilon$ is exactly like the first argument given above. The induction now proceeds along obvious lines, and we omit the remainder of the details, except to note that since $\bigvee\{x_n\} = H$, if M_j has been defined for $j = 1, 2, \ldots,$ then it is certain that γ) is satisfied. This completes the proof of Theorem 10.2, and it follows that the operator $S \oplus M_z^-$ is quasitriangular.

11. Some results of Brown, Douglas and Fillmore

Let us review our position briefly. We have under consideration a connected open set Ω whose boundary consists of the disjoint union of $k \geq 1$ simple closed, smooth Jordan curves $\Gamma_1,\ldots,\Gamma_k$ satisfying conditions a) and b) stated at the beginning of Section 9. Associated with Ω and $\partial\Omega$ are the Hilbert spaces $L_2 = L_2(\partial\Omega)$ and $H_2 = H_2(\partial\Omega)$ defined in Section 9. Furthermore, M_z^+ is the subnormal operator on H_2 having the properties stated in Proposition 9.1, and N is a normal operator acting on the Hilbert space $\mathcal{H}$ and satisfying $\sigma(N) = E(N) = \overline{\Omega}$. Finally, by what has gone before, we know that to prove Theorem A it suffices to show that $N \oplus M_z^+$ is quasitriangular. To this end, let us first observe that by virtue of V) of Proposition 9.1, $N \oplus M_z^+$ has a compact self-commutator. Furthermore, using III) and IV) of Proposition 9.1, we compute easily that $E(N \oplus M_z^+) = W(N \oplus M_z^+) = \overline{\Omega}$. Therefore, the shortest way to complete the proof of Theorem A is simply to quote the following recent remarkable theorem of L. Brown, Douglas and Fillmore announced in [9].

THEOREM B. Let T be any operator in $\mathcal{L}(\mathcal{H})$ such that T has a compact self-commutator and satisfies $E(T) = W(T)$. Then T is the sum of a normal operator and a compact operator.

This theorem shows that $N \oplus M_z^+$ is the sum of a normal operator and a compact operator, and it is known from [14] that all such operators are quasitriangular. Thus the proof of Theorem A might be said to be complete.

The unpleasant feature of completing the proof of Theorem A in this fashion is that Theorem B has considerable depth, and its proof is neither short nor easy. Thus we choose to proceed differently, and give a detailed proof of the fact that $N \oplus M_z^+$ is quasitriangular that does not use Theorem B. (In fact, we shall prove more; namely, that $N \oplus M_z^+$ is in the norm closure of the set of operators that can be written as the sum of a normal operator and a compact operator.) We shall argue by induction on the number of "holes" in the domain Ω, i.e., on the number k of simple closed Jordan curves $\Gamma_1,\ldots,\Gamma_k$ whose union constitutes the boundary of Ω. Before we begin this argument, however, we reproduce (for the sake of completeness) some of the facts from the recent work of Brown, Douglas, and Fillmore [8, 9] that we shall need, and we make a further reduction that allows us to deal with Toeplitz operators.

DEFINITION. We shall say that a compact set Λ in $\mathbb{C}$ has property (BDF) if every operator T in $L(H)$ that has a compact self-commutator and satisfies $E(T) = W(T) = \Lambda$ is the sum of a normal operator and a compact operator. (Observe that $[T,T^*]$ is compact if and only if $\pi(T)$ is normal in the Calkin algebra.)

PROPOSITION 11.1. The unit circle τ has property (BDF).

Proof. It suffices to show that if U is any unitary element of the Calkin algebra $L(H)/\mathbb{K}$, then there is an isometry V in $L(H)$ such that $\pi(V) = U$, and the reader may supply the proof of this fact himself.

PROPOSITION 11.2. Every homeomorphic image Λ of τ has property (BDF).

Proof. It is obvious that any translate of a set with property (BDF) also has property (BDF), so we may assume without loss of generality that Λ contains the origin in its interior. Let ϕ be a homeomorphism of Λ onto τ, and let T be an operator on H such that $\pi(T)$ is normal and such that $E(T) = W(T) = \Lambda$. Clearly $\phi(\pi(T))$ is normal in $L(H)/\mathbb{K}$ and has spectrum τ. Moreover, it suffices to find a normal operator Y on H such that $\pi(Y) = \phi(\pi(T))$, because then $\pi(T) = \phi^{-1}(\pi(Y)) = \pi(\phi^{-1}(Y))$, and thus T is the sum of the normal operator $\phi^{-1}(Y)$ and a compact operator. Let Y_1 be any operator on H such that $\pi(Y_1) = \phi(\pi(T))$. Then, by Proposition 11.1, it suffices to show that $W(Y_1) = \tau$, or equivalently, that $i(Y_1-0) = 0$. On the other hand, we are given that $i(T-0) = 0$, which is equivalent to saying that the invertible element $\pi(T)$ in the Calkin algebra is contained in the connected component of the identity in the group of invertible elements in the Calkin algebra (cf. [11], pp. 35, 36, 138). Thus, consider the commutative C*-algebra A generated by $\pi(T)$, which is C*-isomorphic to the algebra of continuous functions $C(\Lambda)$, and observe that under this isomorphism j, $\pi(T)$ is carried onto the function $z \to z$ in $C(\Lambda)$. Now let g_λ, $0 \leq \lambda \leq 1$, be a continuous arc of functions in $C(\Lambda)$ such that for $z \in \Lambda$, $g_0(z) = z$ and $g_1(z) = \phi(z)$. Then the arc $G_\lambda = j^{-1}(g_\lambda)$, $0 \leq \lambda \leq 1$, in the Calkin algebra has the property that $G_0 = j^{-1}(g_0) = \pi(T)$ and $G_1 = j^{-1}(g_1) = j^{-1}(\phi) = \phi(\pi(T))$ $= \pi(\phi(T)) = \pi(Y_1)$. Since G_0 is in the connected component of the identity in the group of invertible elements in $L(H)/\mathbb{K}$, the element $G_1 = \pi(Y_1)$ must be also, which proves that $i(Y_1) = 0$, and thus that Y_1 is the sum of a normal operator and a compact operator. This completes the proof.

PROPOSITION 11.3. Suppose that Λ_1 and Λ_2 are nonempty, disjoint, compact plane sets each having property (BDF), and suppose that Λ_1 is contained in the unbounded component of $\mathbb{C}\backslash\Lambda_2$. Then $\Lambda_1 \cup \Lambda_2$ has property (BDF).

Proof. We may suppose, by translating if necessary, that $0 \notin \Lambda_1 \cup \Lambda_2$. Let T be an operator on H such that $\pi(T)$ is normal and $E(T) = W(T) = \Lambda_1 \cup \Lambda_2$. By virtue of the hypothesis on Λ_1 and Λ_2, the Riesz functional calculus can be used to construct an idempotent $\pi(F)$ that commutes with $\pi(T)$ and has the property that $E(FTF) = \Lambda_1 \cup \{0\}$ and $E((1-F)T(1-F)) = \Lambda_2 \cup \{0\}$.

Furthermore, since $\pi(T)$ is normal, $\pi(F)$ is a projection, and one knows that in this case, F may be taken to be a projection. We write T as a 2×2 matrix $T = (T_{ij})$ relative to the decomposition $H = FH \oplus (1-F)H$, and observe that since $\pi(TF-FT) = 0$, the matrix entries T_{12} and T_{21} are compact. Thus, to show that T is the sum of a normal operator and a compact operator, it suffices to prove this for $T_{11} \oplus T_{22}$, and, going one step further, it suffices to show that each of T_{11} and T_{22} has this property. Furthermore, since $E(T_{11} \oplus T_{22}) = \Lambda_1 \cup \Lambda_2$, $E(T_{11}) \subset E(T_{11} \oplus 0)$ $= E(FTF) = \Lambda_1 \cup \{0\}$, and similarly $E(T_{22}) \subset \Lambda_2 \cup \{0\}$, it follows th $E(T_{11}) = \Lambda_1$ and $E(T_{22}) = \Lambda_2$. Thus the proof can be completed by showing that $W(T_{11}) = E(T_{11})$ and $W(T_{22}) = E(T_{22})$. To this end, suppose $\lambda \notin \Lambda_1$. If λ belongs to some bounded component of $\mathbb{C} \setminus \Lambda_1$, then λ belongs to the unbounded component of $\mathbb{C} \setminus \Lambda_2$, and thus $i(T_{22}-\lambda) = 0$. Since $i[(T_{11}-\lambda) \oplus (T_{22}-\lambda)] = 0$ by hypothesis, it follows that $i(T_{11}-\lambda) = 0$. On the other hand, if λ belongs to theunbounded component of $\mathbb{C} \setminus \Lambda_1$, then clearly $i(T_{11}-\lambda) = 0$; thus we have established that $W(T_{11}) = E(T_{11})$. It follows easily from this and the fact that $W(T_{11} \oplus T_{22}) = E(T_{11} \oplus T_{22})$ that $W(T_{22}) = E(T_{22})$, so the proof is complete.

COROLLARY 11.4. Let Ω be the domain described at the beginning of this section whose boundary $\partial\Omega$ is equal to $\Gamma_1 \cup \ldots \cup \Gamma_k$. Then the compact set $\partial\Omega$ has property (BDF).

Proof. If $k = 1$, the result follows from Proposition 11.2. If $k > 1$, we use Proposition 11.3, consider first $\Gamma_k \cup \Gamma_{k-1}$, and employ a finite induction argument.

The next result is the deepest fact about sets with property (BDF) that we shall need.

THEOREM 11.5. Suppose that Λ_1 and Λ_2 are homeomorphic images of the unit circle τ such that $\mathrm{Int}(\Lambda_1) \cap \mathrm{Int}(\Lambda_2) = \emptyset$ and such that $\Lambda_1 \cap \Lambda_2$ is a singleton $\{\lambda_0\}$. Then $\Lambda_1 \cup \Lambda_2$ has property (BDF).

Proof. Clearly property (BDF) is invariant under translation of the set by a scalar, so we may and so assume that $\Lambda_1 \cap \Lambda_2 = \{0\}$. Let T be an operator on $\mathcal{H}$ such that $E(T) = W(T) = \Lambda_1 \cup \Lambda_2$ and such that $\pi(T)$ is normal. Our task is to show that T is the sum of a normal operator and a compact operator. The proof will use two elementary facts about elements in an arbitrary C*-algebra $\mathcal{B}$ which the reader can verify for himself:

i) If A and B are positive contractions in $\mathcal{B}$, $A \geq B$, and D is any element in $\mathcal{B}$ such that $BD = D$, then $AD = D$.

ii) If $A \in \mathcal{B}$, B is normal in $\mathcal{B}$, and $AB = 0$, then $AB^* = 0$.

Consider now the commutative C*-subalgebra $\mathcal{A}$ of $\mathcal{L}(\mathcal{H})/\mathbb{K}$ generated by the normal element $\pi(T)$, and observe that there exists a C*-isomorphism j from $\mathcal{A}$ onto $C(\Lambda_1 \cup \Lambda_2)$ such that $j(\pi(T))$ is the function $z \to z$ in $C(\Lambda_1 \cup \Lambda_2)$. We define the function h in $C(\Lambda_1 \cup \Lambda_2)$ by $h(z) = |z|$, $z \in \Lambda_1$, and $h(z) = -|z|$, $z \in \Lambda_2$, and we observe that h vanishes only at the origin. One knows that there exists a Hermitian operator H in $\mathcal{L}(\mathcal{H})$ such that $\pi(H) = j^{-1}(h)$, and by making a compact perturbation of H, if necessary, we may assume that $\sigma(H) = E(H) = \text{range } h = [\alpha,\beta]$, $\alpha < 0 < \beta$, and also that 0 is not in the point spectrum of H. Consider the functions f_1 and f_2 in $C(\Lambda_1 \cup \Lambda_2)$ defined by the equations $f_1(z) = z$, $f_2(z) = 0$, $z \in \Lambda_1$; $f_1(z) = 0$, $f_2(z) = z$, $z \in \Lambda_2$. Then there exist operators T_1 and T_2 on $\mathcal{H}$ such that $\pi(T_i) = j^{-1}(f_i)$, $i = 1,2$, and it is clear from the definitions of f_1 and f_2 that $\pi(T) = \pi(T_1+T_2)$.

We next define two increasing sequences $\{\phi_n\}$ and $\{\psi_n\}$ in $C([\alpha,\beta])$ as follows. For each positive integer n, $\phi_n(t) \equiv 1$ on $[\alpha,-\frac{1}{n}]$, $\phi_n(t) \equiv 0$ on $[-\frac{1}{2n}, \beta]$, and ϕ_n is linear on $[-\frac{1}{n},-\frac{1}{2n}]$. Also $\psi_n(t) \equiv 1$ on $[\frac{1}{n}, \beta]$, $\psi_n(t) \equiv 0$ on $[\alpha, \frac{1}{2n}]$, and ψ_n is linear on $[\frac{1}{2n}, \frac{1}{n}]$. Clearly $\{\phi_n\}$ converges pointwise to $\chi_{[\alpha,0)}$ and $\{\psi_n\}$ converges pointwise to $\chi_{(0,\beta]}$. Moreover, it is an easy consequence of the spectral theorem that the increasing sequences $\{\phi_n(H)\}$ and $\{\psi_n(H)\}$ of positive contractions converge weakly to their least upper bounds $F = \chi_{[\alpha,0)}(H)$ and $G = \chi_{(0,\beta]}(H)$, respectively. Furthermore it is obvious that F and G are projections, and since 0 is not in the point spectrum of H , $G = 1-F$. Since π is order preserving, for every positive integer n we have

$$\pi(F) \geq \pi(\phi_n(H)) = \phi_n(\pi(H)) = \phi_n(j^{-1}(h)) = j^{-1}(\phi_n(h)) ,$$

and similarly, $\pi(1-F) \geq j^{-1}(\psi_n(h))$.

We prove next that $\pi(F)\pi(T_1) = \pi(T_1)$. For this purpose, let g be any function in $C(\Lambda_1 \cup \Lambda_2)$ that vanishes on Λ_2 and also on some open set containing 0 . Then for n sufficiently large, $\phi_n(h)g = g$, and thus $j^{-1}(\phi_n(h))j^{-1}(g) = j^{-1}(g)$. From i) we obtain $\pi(F)j^{-1}(g) = j^{-1}(g)$, and since f_1 can be approximated uniformly by such functions g , it follows that $\pi(F)j^{-1}(f_1) = j^{-1}(f_1)$, or, equivalently, $\pi(F)\pi(T_1) = \pi(T_1)$. An analogous argument shows that $\pi(1-F)\pi(T_2) = \pi(T_2)$,or, equivalently, that $\pi(F)\pi(T_2) = 0$. Since $\pi(T_1)$ and $\pi(T_2)$ are normal, it follows from ii) that

$$\pi(FT_1) = \pi(T_1F) = \pi(FT_1F) = \pi(T_1)$$

and

$$\pi(FT_2) = \pi(T_2F) = \pi(FT_2F) = 0 .$$

Thus

$$\pi(T_1) = \pi(FT_1) = \pi(FT) = \pi(TF) = \pi(FTF) ,$$

and in particular, FT-TF is compact.

We now write T as a 2×2 matrix (T_{ij}) relative to the decomposition $\mathcal{H} = F\mathcal{H} \oplus (1-F)\mathcal{H}$, and the fact that FT-TF is compact implies that T is the sum of a compact operator and $T_{11} \oplus T_{22}$. Thus to complete the proof, it suffices to show that each of T_{11} and T_{22} is the sum of a normal operator and a compact operator (on its own Hilbert space), and we argue as follows. Observe that

$$\Lambda_1 = E(T_1) = E(FT_1F) = E(FTF) = E(T_{11}\oplus 0) = E(T_{11}) \cup \{0\} .$$

Since Λ_1 is perfect, it follows that $E(T_{11}) = \Lambda_1$, and an analogous argument shows that $E(T_{22}) = \Lambda_2$. Thus, by Proposition 11.2, it suffices to show that $W(T_{11}) = \Lambda_1$ and $W(T_{22}) = \Lambda_2$; by symmetry it suffices to show that if $\lambda \in \mathrm{Int}(\Lambda_1)$, then $i(T_{11}-\lambda) = 0$. Since $\pi(T_1) = \pi(FTF) = \pi(T_{11}\oplus 0)$, this is equivalent to proving that $i(T_1-\lambda) = 0$, or, what is the same thing, that $\pi(T_1-\lambda)$ belongs to the connected component G_0 of the identity in the group of invertible elements in $L(H)/\mathbb{K}$. On the other hand we are given that $i(T-\lambda) = 0$, which implies that $\pi(T-\lambda)$ belongs to G_0. Since the function $z-\lambda$ is homotopically equivalent to the function $f_1(z)-\lambda$ in $C(\Lambda_1 \cup \Lambda_2)$, it follows easily that there exists a continuous arc of invertible elements in $L(H)/\mathbb{K}$ joining $\pi(T-\lambda)$ to $\pi(T_1-\lambda)$, which shows that $i(T_1-\lambda) = 0$ and completes the proof.

12. A reduction involving Toeplitz operators

We know that to complete the proof of Theorem A, it suffices to show that the operator $N \oplus M_z^+$ described at the beginning of Section 11 is quasitriangular. To accomplish this, we shall make use of some elementary properties of certain Toeplitz operators, and we begin with a brief review of the pertinent facts. Let L^2 be the Hilbert space consisting of all complex functions f defined on the unit circle τ and square integrable with respect to normalized Lebesgue measure on τ. Furthermore, let H^2 be the subspace of L^2 generated by the orthonormal sequence of vectors $\{e^{int}\}_{n=0}^{\infty}$ in L^2. As usual, we write $C(\tau)$ for the algebra of continuous, complex valued functions on τ. If $\phi \in C(\tau)$, then the <u>Laurent operator</u> L_ϕ is the multiplication operator on L^2 defined by $L_\phi f = \phi f$, and the <u>Toeplitz operator</u> T_ϕ is the operator on H^2 defined by $T_\phi f = PL_\phi f$, where P is the projection of L^2 onto H^2. If Γ is a simple closed Jordan curve, and if ϕ is a homeomorphism of τ onto Γ such that each point λ_0 in the bounded component of $\mathbb{C}\backslash\Gamma$ has index (i.e., winding number) +1 relative to ϕ, we denote by $\tilde{\phi} : e^{it} \to \phi(e^{-it})$ the associated homeomorphism of τ onto Γ with the property that each point λ_0 in the bounded component $\mathbb{C}\backslash\Gamma$ has index -1 relative to $\tilde{\phi}$. <u>In the remainder of the paper we adopt the convention that every homeomorphism of</u> τ <u>(into $\mathbb{C}$) that is written without [resp., with] a "tilde" has the property that the index of any point</u> λ_0 <u>in the bounded component of the complement of the associated simple closed Jordan curve has index</u> +1 [<u>resp</u>., -1].

It is known (cf. [11, Chapter 7]) that if ϕ is a homeomorphism of τ into $\mathbb{C}$ and $\lambda_0 \in \mathrm{Int}(\phi(\tau))$, then the Toeplitz operators $T_\phi - \lambda_0$ and $T_{\tilde{\phi}} - \lambda_0$ are Fredholm operators with indices -1 and +1, respectively. Furthermore, $E(T_\phi) = E(T_{\tilde{\phi}})$, $\sigma(T_\phi) = \sigma(T_{\tilde{\phi}}) = \phi(T) \cup \mathrm{Int}(\phi(T))$, and both T_ϕ and $T_{\tilde{\phi}}$ have compact self-commutators.

PROPOSITION 12.1. Let Γ be a simple closed Jordan curve, and let ϕ be a homeomorphism of τ onto Γ. Then the operator $T_\phi \oplus T_{\tilde{\phi}}$ is unitarily equivalent to a compact perturbation of L_ϕ [resp., $L_{\tilde{\phi}}$].

Proof. It follows easily from what was said above that $E(T_\phi \oplus T_{\tilde{\phi}}) = W(T_\phi \oplus T_{\tilde{\phi}}) = \Gamma$, and since $E(L_\phi) = E(L_{\tilde{\phi}}) = \Gamma$, the result follows from Proposition 11.2 and the converse of the Berg-von Neumann-Weyl theorem.

As was mentioned earlier, we are going to demonstrate the quasi-triangularity of the operator $N \oplus M_z^+$ (described at the beginning of Section 11) by proving a stronger result. In this connection, the following notation will be useful. If K is any separable Hilbert space, we denote by $(N+C) = (N+C)_K$ the set of all operators on K that can be written as the sum of a normal operator on K and a compact operator on K . Moreover, if A is any operator on K , we will write $d(A) = d_K(A)$ for the metric distance

$$d_K(A) = \inf_{T \in (N+C)_K} ||A - T||$$

from A to the set $(N+C)$. For the sake of convenience, since there will be several Hilbert spaces under consideration at one time, we allow ourselves to use the same notation $d(\ )$ without the Hilbert space subscripts, since no confusion can possibly result from this practice. We note for future use the fact that the function d has the property that $d(A_1 \oplus \ldots \oplus A_k) \leq \max_{1 \leq i \leq k} d(A_i)$.

Since every operator in $(N+C)$ is known to be quasitriangular [14], and since the set of quasitriangular operators is closed in the norm topology [14], we can prove that $N \oplus M_z^+$ is quasitriangular by showing that $d(N \oplus M_z^+) = 0$, and this we shall do.

If A and B are operators, we shall find it convenient, henceforth, to write $A \sim B$ to mean that A is unitarily equivalent to a compact perturbation of B . Note the obvious fact that if $A \sim B$, then $d(A) = 0$ if and only if $d(B) = 0$.

The following proposition allows us to replace the problem of proving that $d(N \oplus M_z^+) = 0$ with an analogous problem involving Toeplitz operators.

PROPOSITION 12.2. Let N , M_z^+ , Ω , and $\Gamma_1, \ldots, \Gamma_k$ be as described at the beginning of Section 11, and let ϕ_1 , $\tilde{\phi}_2, \ldots, \tilde{\phi}_k$ be homeomorphisms of τ onto $\Gamma_1, \ldots, \Gamma_k$, respectively, with the indicated orientations. Then $d(N \oplus M_z^+) = 0$ if $d(N \oplus T_{\phi_1} \oplus T_{\tilde{\phi}_2} \oplus \ldots \oplus T_{\tilde{\phi}_k}) = 0$. (Of course, if $k = 1$ this last operator is simply $N \oplus T_{\phi_1}$.)

Proof. Since $E(N) = \overline{\Omega} \supset \Gamma_1 \cup \ldots \cup \Gamma_k = E(L_{\phi_1}) \cup \ldots \cup E(L_{\tilde{\phi}_k})$, it follows from the converse of the Berg-von Neumann-Weyl theorem that $N \sim N \oplus L_{\phi_1} \oplus \ldots \oplus L_{\tilde{\phi}_k}$. Furthermore, by applying Proposition 12.1, we obtain

$$N \sim N \oplus (T_{\phi_1} \oplus T_{\tilde{\phi}_1}) \oplus \ldots \oplus (T_{\phi_k} \oplus T_{\tilde{\phi}_k}) .$$

On the other hand, if we denote by A the direct sum

$$A = M_z^+ \oplus T_{\tilde{\phi}_1} \oplus T_{\phi_2} \oplus \ldots \oplus T_{\phi_k} .$$

then an elementary calculation together with Proposition 9.1 discloses that $E(A) = W(A) = \partial\Omega$ and also that A has a compact self-commutator. Thus, by Corollary 11.4, A is the sum of a normal operator and a compact operator. Since

$$N \oplus M_z^+ \sim (N \oplus T_{\phi_1} \oplus T_{\tilde{\phi}_2} \oplus \ldots \oplus T_{\tilde{\phi}_k}) \oplus A$$

the result follows.

The plan to prove that the operator $D = N \oplus T_{\phi_1} \oplus \ldots \oplus T_{\tilde{\phi}_k}$ of Proposition 12.2 satisfies $d(D) = 0$ is to show that for every positive number ε , $d(D) < \varepsilon$. The following theorem contains the main idea involved in this approximation argument.

THEOREM 12.3. Let N and Ω be as described at the beginning of Section 11, let ε be positive, and let $\psi_1, \ldots, \psi_n$ be homeomorphisms of τ such that $\psi_1(\tau) \cup \ldots \cup \psi_n(\tau) \subset \overline{\Omega}$ and such that for $1 \leq i \leq n-1$, $||\psi_{i+1} - \psi_i||_{C(\tau)} < \varepsilon$. Then,

$$d(N \oplus T_{\psi_1}) \leq d(N \oplus T_{\psi_n}) + \varepsilon ,$$

and, more generally, if S is any operator,

$$d(N \oplus T_{\psi_1} \oplus S) \leq d(N \oplus T_{\psi_n} \oplus S) + \varepsilon .$$

Proof. Since $\psi_1(\tau) \cup \ldots \cup \psi_n(\tau) \subset \overline{\Omega} = E(N)$, it follows as before that

$$N \sim (N \oplus L_{\psi_1} \oplus \ldots \oplus L_{\psi_n}) \sim [N \oplus (T_{\psi_1} \oplus T_{\tilde{\psi}_1}) \oplus \ldots \oplus (T_{\psi_n} \oplus T_{\tilde{\psi}_n})] ,$$

and hence

$$(N\oplus T_{\psi_1}) \sim [(N\oplus T_{\psi_1} \oplus T_{\tilde{\psi}_1}) \oplus (T_{\psi_1} \oplus T_{\tilde{\psi}_2}) \oplus \ldots \oplus (T_{\psi_{n-1}} \oplus T_{\tilde{\psi}_n}) \oplus T_{\psi_n}]$$

Moreover, for $1 \leq i \leq n-1$,

$$||(T_{\psi_i} \oplus T_{\tilde{\psi}_{i+1}}) - (T_{\psi_{i+1}} \oplus T_{\tilde{\psi}_{i+1}})|| \leq ||T_{\psi_{i+1}} - T_{\psi_i}|| \leq ||\psi_{i+1} - \psi_i||_{C(\tau)} < \varepsilon .$$

and each operator $T_{\psi_{i+1}} \oplus T_{\tilde{\psi}_{i+1}}$ satisfies $d(T_{\psi_{i+1}} \oplus T_{\tilde{\psi}_{i+1}}) = 0$, since it is unitarily equivalent to a compact perturbation of the normal operator $L_{\psi_{i+1}}$ (Proposition 12.1). That $d(N\oplus T_{\psi_1}) \leq d(N\oplus T_{\psi_n}) + \varepsilon$ now follows from the fact, mentioned earlier, that $d(A_1 \oplus \ldots \oplus A_m) \leq \max_{1\leq i\leq m} d(A_i)$. The argument which proves that for any operator S , $d(N\oplus T_{\psi_1} \oplus S) \leq d(N\oplus T_{\psi_n} \oplus S) + \varepsilon$, is essentially the same, and we omit it.

13. The coup de grace.

In this section we complete the proof of Theorem A by establishing the following result.

THEOREM 13.1. Let $k \geq 1$, and let the operator N , the domain Ω, the simple closed Jordan curves $\Gamma_1,\ldots,\Gamma_k$, and the homeomorphisms $\phi_1,\ldots,\phi_k$ be as in Proposition 12.2. Furthermore, let $D = D_k$ be the operator $N \oplus T_{\phi_1} \oplus T_{\tilde{\phi}_2} \oplus \ldots \oplus T_{\tilde{\phi}_k}$ (where, once again, it is understood that if $k = 1$, then $D = N \oplus T_{\phi_1}$). Then $d(D) = 0$ (and hence $d(N\oplus M_z^+) = 0$).

Proof. The proof is by induction on the integer k . Suppose first that $k = 1$. Then $\partial\Omega = \Gamma_1$, $\overline{\Omega} = \Gamma_1 \cup \mathrm{Int}(\Gamma_1)$ is simply connected, and $D = N \oplus T_{\phi_1}$. Let $\varepsilon > 0$ be given, and recall that it suffices to prove that $d(N\oplus T_{\phi_1}) < \varepsilon$. To accomplish this, let λ_0 be a point in Ω , and let $\psi_1 = \phi_1, \psi_2, \ldots, \psi_n$ be a sequence of homeomorphisms of τ all of whose ranges are contained in $\overline{\Omega}$ and having the additional properties that

$$||\psi_{i+1}-\psi_i||_{C(\tau)} < \varepsilon/2,\ 1 \le i \le n-1, \text{ and } ||\psi_n-\lambda_0||_{C(\tau)} < \varepsilon/2 .$$

(That such a sequence of homeomorphisms exists is an immediate consequence of the contractibility of $\overline{\Omega}$.) Then, by Theorem 12.3, $d(N\oplus T_{\phi_1}) \le d(N\oplus T_{\psi_n}) + \varepsilon/2 \le d(N\oplus\lambda_0) + \varepsilon = \varepsilon$, which completes the proof in the case $k = 1$. The following diagram, where $\gamma_i = \psi_i(\tau)$, gives an intuitive view of the argument.

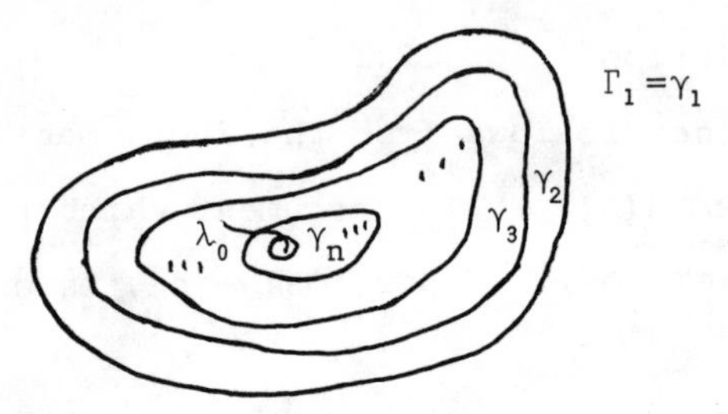

Suppose now that $k = 2$. Then Ω is an "annulus like" domain whose boundary $\partial\Omega$ equals $\Gamma_1 \cup \Gamma_2$, and we wish to show that $d(D) = 0$, where $D = N \oplus T_{\phi_1} \oplus T_{\tilde{\phi}_2}$. Let $\varepsilon > 0$ be given. Once again, it is easy topology that there exists a sequence $\psi_1 = \phi_1$, $\psi_2,\ldots,\psi_m = \phi_2$ of homeomorphisms of τ into $\overline{\Omega}$ such that for $1 \le i \le m-1$, $||\psi_{i+1}-\psi_i||_{C(\tau)} < \varepsilon$. Thus by Theorem 12.3 (with $S = T_{\tilde{\phi}_2}$) ,

$$d(N\oplus T_{\phi_1}\oplus T_{\tilde{\phi}_2}) \le d(N\oplus T_{\psi_m}\oplus T_{\tilde{\phi}_2}) + \varepsilon ,$$

and since $T_{\psi_m} \oplus T_{\tilde{\phi}_2} = T_{\psi_m} \oplus T_{\tilde{\psi}_m} \sim L_{\psi_m}$, which belongs to $(N+C)$, we have $d(N\oplus T_{\phi_1}\oplus T_{\tilde{\phi}_2}) \le d(N\oplus L_{\psi_m}) + \varepsilon = \varepsilon$, which completes the proof in case $k = 2$. The following diagram, with $\gamma_i = \psi_i(\tau)$ gives an intuitive view of the argument.

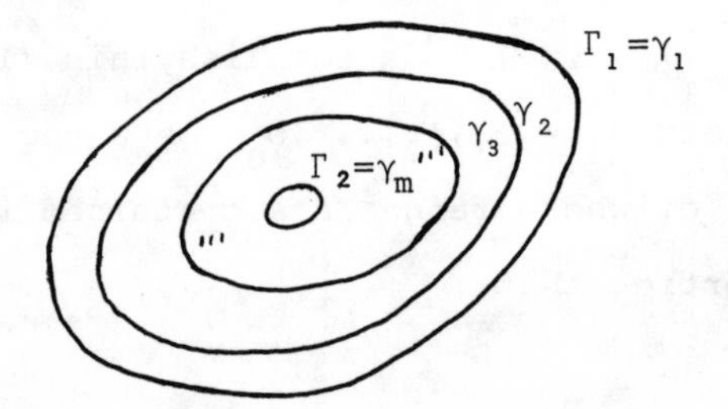

Turning now to the inductive step, we suppose the theorem to be true in all cases where the boundary of the domain Ω consists of $n \leq k$ simple closed Jordan curves with the appropriate properties, and we now consider the case that $\partial\Omega = \Gamma_1 \cup \ldots \cup \Gamma_{k+1}$. We wish to show that the operator $D = N \oplus T_{\phi_1} \oplus T_{\tilde{\phi}_2} \oplus \ldots \oplus T_{\tilde{\phi}_{k+1}}$ satisfies $d(D) = 0$; to accomplish this, we let ε be a given positive number and prove that $d(D) < \varepsilon$. For this purpose, we make the following construction. Let Λ_1 and Λ_2 be simple closed Jordan curves with the following properties: $\Lambda = \Lambda_1 \cup \Lambda_2 \subset \Omega$, $\Gamma_2 \subset \mathrm{Int}(\Lambda_1)$, $\Gamma_3 \cup \ldots \cup \Gamma_{k+1} \subset \mathrm{Int}(\Lambda_2)$, and $\Lambda_1 \cap \Lambda_2 = \{\lambda_0\}$. Furthermore, let δ be a continuous function on τ such that $\delta(\tau) = \Lambda$, such that δ "traces out" Λ exactly once, and such that every point λ in $\mathrm{Int}(\Lambda_1) \cup \mathrm{Int}(\Lambda_2)$ has index +1 relative to δ. In keeping with the notation for homeomorphisms, we define $\tilde{\delta}$ by $\tilde{\delta}(e^{it}) = \delta(e^{-it})$. We shall also need homeomorphisms δ_1 and δ_2 of τ such that $\delta_1(\tau) = \Lambda_1$ and $\delta_2(\tau) = \Lambda_2$. Finally, let N_1 and N_2 be normal operators such that

$$\sigma(N_1) = E(N_1) = \Lambda_1 \cup \Gamma_2 \cup [\mathrm{Int}(\Lambda_1) \cap \mathrm{Ext}(\Gamma_2)]$$

and such that

$$\sigma(N_2) = E(N_2) = \Lambda_2 \cup \Gamma_3 \cup \ldots \cup \Gamma_{k+1} \cup [\mathrm{Int}(\Lambda_2) \cap \mathrm{Ext}(\Gamma_3) \cap \ldots \cap \mathrm{Ext}(\Gamma_{k+1})] .$$

Observe that by the converse to the Berg-von Neumann-Weyl theorem we have that $N \sim N \oplus N_1 \oplus N_2 \oplus L_\delta$, and by virtue of Theorem 11.5 and some obvious calculations, we have $L_\delta \sim T_\delta \oplus T_{\tilde{\delta}_1} \oplus T_{\tilde{\delta}_2}$. Thus

$$\begin{aligned} D &\sim (D \oplus N_1 \oplus N_2 \oplus L_\delta) \sim (D \oplus N_1 \oplus N_2 \oplus T_\delta \oplus T_{\tilde{\delta}_1} \oplus T_{\tilde{\delta}_2}) \\ &\sim (N \oplus T_{\phi_1} \oplus T_{\tilde{\delta}}) \oplus (N_1 \oplus T_{\delta_1} \oplus T_{\tilde{\phi}_2}) \oplus (N_2 \oplus T_{\delta_2} \oplus T_{\tilde{\phi}_2} \oplus \ldots \oplus T_{\tilde{\phi}_{k+1}}) . \end{aligned}$$

By virtue of the induction hypothesis, the operators inside the second and third pairs of parentheses above are at distance zero from $(N+C)$, so to complete the proof of the theorem, it suffices to show that $d(N\oplus T_{\phi_1}\oplus T_{\tilde{\delta}}) < \varepsilon$. We argue as in the case $k = 2$. Namely, we define a sequence $\psi_1 = \phi_1, \psi_2, \ldots, \psi_n$ of homeomorphisms of τ into $\overline{\Omega} \cap (\mathrm{Ext}(\Lambda_1) \cup \mathrm{Ext}(\Lambda_2))$ such that for $1 \leq i \leq n-1$, $||\psi_{i+1}-\psi_i||_{C(\tau)} < \varepsilon$, and such that $||\psi_n-\delta||_{C(\tau)} < \varepsilon$. Then, by Theorem 12.3, $d(N\oplus T_{\phi_1}\oplus T_{\tilde{\delta}}) \leq d(N\oplus T_{\psi_n}\oplus T_{\tilde{\delta}}) + \varepsilon$ and since

$$||(T_{\psi_n}\oplus T_{\tilde{\delta}}) - (T_{\psi_n}\oplus T_{\tilde{\psi}_n})|| \leq ||T_{\tilde{\delta}} - T_{\tilde{\psi}_n}|| \leq ||\tilde{\delta} - \tilde{\psi}_n||_{C(\tau)} < \varepsilon$$

and $(T_{\psi_n}\oplus T_{\tilde{\psi}_n}) \sim L_{\psi_n}$, we have $d(N\oplus T_{\phi_1}\oplus T_{\tilde{\delta}}) < 2\varepsilon$, which completes the proof. The intuitive diagram in this case looks like this, where $\gamma_i = \psi_i(\tau)$:

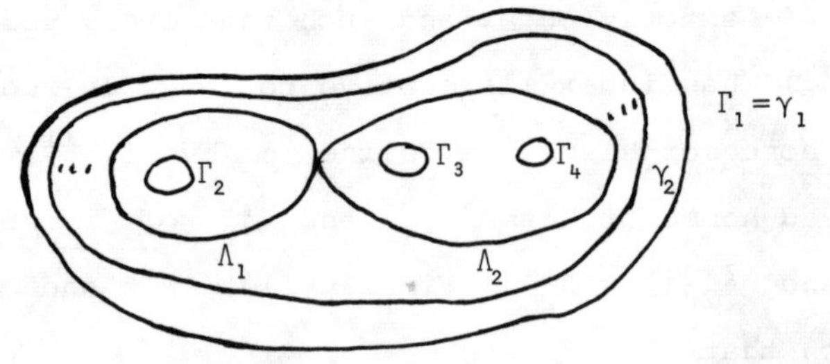

This completes our proof of Theorem A, based only on the results of [13] and [14], and it is our hope that this complete account will be helpful to the understanding of this beautiful theorem of Apostol, Foiaş and Voiculescu.

State University of New York at Stony Brook
University of Michigan

References

[1] C. Apostol, Quasitriangularity in Hilbert space, Indiana U. Math. J., 1973.

[2] C. Apostol, C. Foiaş, and L. Zsido, Some results on nonquasitriangular operators, Indiana U. Math. J., 1973.

[3] C. Apostol, C. Foiaş, and D. Voiculescu, Some results on nonquasitriangular operators II, Rev. Roum. Math. Pures et Appl., 1973.

[4] ______________, Some results on nonquasitriangular operators III, Rev. Roum. Math. Pures et Appl., 1973.

[5] ______________, Some results on nonquasitriangular operators IV, Rev. Roum. Math. Pures et Appl., 1973.

[6] C. Berger and B. Shaw, Self-commutators of multicyclic hyponormal operators are always trace class, to appear in Bull. Amer. Math. Soc. 79 (1973).

[7] A. Brown and C. Pearcy, Compact restrictions of operators, Acta Sci. Math. (Szeged), 32 (1971) 271-282.

[8] L. Brown, R. Douglas and P. Fillmore, Unitary equivalence modulo the compact operators and extensions of C*-algebras, these Notes.

[9] L. Brown, R. Douglas and P. Fillmore, Extensions of C*-algebras, operators with compact self-commutators, and K-homology, to appear in Bull. Amer. Math. Soc. 79 (1973).

[10] D. Deckard, R. Douglas and C. Pearcy, On invariant subspaces of quasitriangular operators, Amer. J. Math., 91 (1969) 637-647.

[11] R. Douglas, <u>Banach Algebra Techniques in Operator Theory</u>, Academic Press, New York, 1972.

[12] R. Douglas, <u>Banach Algebra Techniques in the Theory of Toeplitz Operators</u>, CBMS Lecture 15, American Math. Soc. Providence, 1973.

[13] R. Douglas and C. Pearcy, A note on quasitriangular operators, Duke Math. J., 37 (1970) 177-188.

[14] P. R. Halmos, Quasitriangular operators, Acta Sci. Math. (Szeged), 29 (1968) 283-293.

[15] C. Pearcy, Some unsolved problems in operator theory, preprint circulated in 1972.

[16] C. Pearcy and N. Salinas, An invariant subspace theorem, Mich. Math. J., 20 (1973) 21-31.

[17] C. Pearcy and N. Salinas, Compact perturbations of semi-normal operators, Indiana U. Math. J., 22 (1973) 789-793.

[18] D. Sarason, Representing measures for R(X) and their Greens functions, J. Functional Anal., 7 (1971) 359-385.

UNITARY EQUIVALENCE MODULO THE COMPACT OPERATORS

AND

EXTENSIONS OF C*-ALGEBRAS

L. G. Brown, R. G. Douglas* and P. A. Fillmore

§1. Introduction

Throughout the paper, H will denote a separable complex Hilbert space, $L(H)$ the algebra of bounded linear operators on H, $K(H)$ the ideal of compact operators on H, $A(H)$ the Calkin algebra $L(H)/K(H)$, and $\pi : L(H) \to A(H)$ the quotient map.

We begin with the following question: for operators S,T in $L(H)$, when is S unitarily equivalent to some compact perturbation of T? The question is suggested - and answered for self-adjoint operators - by familiar results of Weyl and von Neumann. In 1909 Weyl proved that if S and K are self-adjoint and K is compact, then the spectra of S and $S+K$ have the same "limit points" (these are the topological limit points together with the isolated eigenvalues of infinite multiplicity), and in 1935 von Neumann supplied a striking converse: if the spectra of the self-adjoint operators S and T have the same limit points, then there is a compact operator K such that $S+K$ and T are unitarily equivalent. Thus, in this case, the question of unitary equivalence modulo the compacts is determined entirely by the set of limit points of the spectrum.

What about normal operators? In proving his result, von Neumann first reduced it to the case of diagonal operators (those possessing an orthonormal basis of eigenvectors) by using the fact, also due to

---

* Sloan Foundation Fellow.

Weyl, that any self-adjoint operator can be compactly perturbed so as to be diagonal. Since Berg [5] has shown that this result is valid for normal operators, and since the rest of von Neumann's argument applies without change, the situation is the same for normal operators.

Before going on, notice that our question is precisely the question of unitary equivalence in the Calkin algebra A. Of course, the spectrum relative to A (the essential spectrum) is a unitary invariant. On the other hand, when N is normal the spectrum of $\pi(N)$ is just the set of limit points of the spectrum of N [21, Th. 3.4], and hence the spectrum is a "complete" unitary invariant for elements of the Calkin algebra that are determined by normal operators.

In this paper we will be concerned with operators T that determine normal elements of the Calkin algebra (that is, $\pi(T)$ is normal, or equivalently, $T^*T - TT^*$ is compact). Such operators will be called _essentially normal_. One way in which these operators arise is as compact perturbations of normals, and to these the foregoing analysis applies. The interest in essentially normal operators is due to the fact that not all of them arise in this fashion.

1.1 Example. If S is the unilateral shift operator, then S is essentially normal, and yet it is not true that $S = N+K$ with N normal and K compact. This is because S is Fredholm of index -1, and it would follow that N is likewise Fredholm of index -1, contradicting the fact that a normal Fredholm operator has index 0 .

(We pause to remark that the example raises the possibility of two notions of unitary equivalence in A, the stronger requiring unitary elements that arise from unitary (equivalently, normal) operators, the other allowing unitaries that do not, such as $\pi(S)$. We show in Theorem 4.3 and Corollary 4.4 that these notions coincide.)

The next example shows that the spectrum is not a complete unitary invariant for normal elements of A.

<u>1.2 Example</u>. If S is the unilateral shift and B the bilateral shift, then both $\pi(S)$ and $\pi(B)$ are normal with spectrum the unit circle. Suppose that $\pi(B)\pi(T) = \pi(T)\pi(S)$ for some unitary (or merely invertible) element $\pi(T)$ of A, so that $BT = TS+K$ with K compact. Since $\pi(T)$ is invertible, Atkinson's Theorem [4] asserts that T is Fredholm. Using the fact that a product of Fredholm operators is Fredholm with index the sum of the indices, we obtain

$$0 = \text{ind } B = \text{ind } S = -1 \quad ,$$

a contradiction.

As this example suggests, there are further unitary invariants arising from index theory. For example, it will be shown (Theorem 3.1) that if $\pi(T)$ is unitary, then T is determined up to unitary equivalence modulo the compacts by the spectrum of $\pi(T)$ and the index of T. In particular, any isometry of finite defect n is a compact perturbation of a shift of multiplicity n.

We turn now to extensions. If X is a compact Hausdorff (usually metric) space, then an extension of K by $C(X)$ is a pair (E,ϕ), where E is a C*-subalgebra of $L(H)$ that contains K and the identity operator I, and ϕ is a *-homomorphism of E onto $C(X)$ with kernel K. The question we wish to consider is the classification, in a sense to be made precise, of such extensions. This problem is closely related to the one already introduced. Notice, for example, that any T in E is essentially normal (since $\phi(T^*T-TT^*) = 0$).

To be more precise, let Λ be a compact subset of the plane, let (E,ϕ) be an extension of K by $C(\Lambda)$, let χ be the identity function $\chi(z) = z$ in $C(\Lambda)$, and fix T in E with $\phi(T) = \chi$. Then E is generated by T, I and K. For if E' is the C*-subalgebra generated by these operators, then $\phi(E')$ is a

C*-subalgebra of $C(\Lambda)$ that contains χ, and thus is all of $C(\Lambda)$. If then S is in $\mathcal{E}$, there exists S' in $\mathcal{E}'$ such that $\phi(S) = \phi(S')$, so that $\phi(S-S') = 0$, $S-S'$ is compact, and hence S is in $\mathcal{E}'$.

On the other hand, let T be any essentially normal operator, let Λ be the spectrum of $\pi(T)$, and let $\mathcal{E}$ be the C*-algebra generated by T, I and $\mathcal{K}$. Then there is a *-homomorphism ϕ of $\mathcal{E}$ onto $C(\Lambda)$ with kernel $\mathcal{K}$ such that $\phi(T) = \chi$. For by the spectral theorem (applied in a faithful representation of $\mathcal{A}$) there is a *-isomorphism ψ from $\pi(\mathcal{E})$ onto $C(\Lambda)$ with $\psi\pi(T) = \chi$, and we can take $\phi = \psi\pi$.

These facts demonstrate the connection between essentially normal operators and extensions of $\mathcal{K}$ by $C(\Lambda)$ with Λ a compact subset of the plane. As to the connection between the classification problems, if we agree that extensions $(\mathcal{E}_1, \phi_1)$ and $(\mathcal{E}_2, \phi_2)$ of $\mathcal{K}$ by $C(X)$ are equivalent if there is a unitary operator U such that $U^*\mathcal{E}_2 U = \mathcal{E}_1$ and $\phi_2(T) = \phi_1(U^*TU)$ for all T in $\mathcal{E}_2$, then it follows that two extensions, generated as above by operators T_1 and T_2 such that $\pi(T_1)$ and $\pi(T_2)$ have the same spectrum, are equivalent if and only if the operators are unitarily equivalent modulo the compacts. For assuming the extensions equivalent we have

$$\phi_1(U^*T_2U) = \phi_2(T_2) = \chi = \phi_1(T_1)$$

so that $U^*T_2U - T_1$ is compact; conversely, if $U^*T_2U - T_1$ is compact then clearly $U^*\mathcal{E}_2U = \mathcal{E}_1$ (by continuity, since $T \to U^*TU$ carries a generating set to a generating set) and $\phi_2(T) = \phi_1(U^*TU)$ for all T in $\mathcal{E}_2$ (for the same reasons).

Thus the classification problems for essentially normal operators and for extensions of $\mathcal{K}$ by $C(\Lambda)$ with Λ a compact subset of the plane are identical. The extension point of view has a number of advantages. To begin with, it suggests the more general problem in which Λ is replaced by any compact Hausdorff space. Next, we have the useful fact that a classification of extensions by a given $C(X)$ amounts to a classification of extensions by $C(Y)$ for any Y homeomorphic to X (because $C(X)$ and $C(Y)$ are *-isomorphic). In particular, if Λ_1 and Λ_2 are homeomorphic subsets of the plane, and operators T such that $\pi(T)$ is normal with spectrum Λ_1 have been classified, then so have operators S such that $\pi(S)$ is normal with spectrum Λ_2. We illustrate this with a final example.

<u>1.3 Example</u>. If $\pi(T)$ is normal with spectrum Λ lying on a simple arc, then T is a compact perturbation of a normal operator. To see this let $\Delta \subset \mathbb{R}$ be homeomorphic to Λ. If S is any operator such that $\pi(S)$ is normal with spectrum Δ, then $\pi(S)$ is self-adjoint, $\pi(S-S^*) = 0$, $S = \operatorname{Re} S +$ compact, and hence any two such operators are unitarily equivalent modulo the compacts by the Weyl-von Neumann Theorem.

Now let E_1 be the C*-algebra generated by I, T and K, let E_2 be the C*-algebra generated by I, K and a normal operator N such that $\pi(N)$ has spectrum Λ, and let (E_1,ϕ_1) and (E_2,ϕ_2) be the extensions of K by $C(\Lambda)$ such that $\phi_1(T) = \chi = \phi_2(N)$. If η denotes a homeomorphism between Δ and Λ, and $\eta^* : C(\Lambda) \to C(\Delta)$ is the dual *-isomorphism $(\eta^*(f) = f\circ\eta)$, then $(E_1,\eta^*\phi_1)$ and $(E_2,\eta^*\phi_2)$ are extensions of K by $C(\Delta)$ and hence are equivalent. If the unitary operator U implements this equivalence, then

$$\eta^*\phi_1(U^*NU) = \eta^*\phi_2(N) = \eta^*(\chi) = \eta^*\phi_1(T)$$

and therefore $U^*NU - T$ is compact as asserted.

Equivalently, we can apply the continuous function η^{-1} to the normal element $\pi(T)$ to obtain a self-adjoint element $\eta^{-1}(\pi(T))$. As in the first paragraph of the example, there exists a self-adjoint operator H such that $\pi(H) = \eta^{-1}(\pi(T))$; moreover we can suppose that H has spectrum Δ. Then $N = \eta(H)$ is normal and

$$\pi(N) = \pi\eta(H) = \eta(\pi(H)) = \pi(T)$$

so T is a compact perturbation of N. Although this is merely another way of looking at the same proof, the use of the functional calculus in $\mathcal{A}$ is natural in the context of extensions.

That an essentially normal operator T with spectrum lying on a simple arc is a compact perturbation of a normal operator was obtained independently by Lancaster [27].

A further advantage of the extension formulation is that it permits the introduction of language and techniques from homological algebra. Denote by $\mathrm{Ext}(X)$ the set of equivalence classes of extensions of $\mathcal{K}$ by $C(X)$. Then $\mathrm{Ext}(X)$ admits a natural commutative binary operation. An obvious candidate for the identity element is the class of trivial (that is, split) extensions. For X metric we show (Theorems 5.3 and 5.5) that all trivial extensions are equivalent (a result that extends the Weyl-von Neumann-Berg Theorem), and that this class acts as the identity element (this result contains the fact, mentioned above, that isometries are compact perturbations of shifts). Further, we show that $\mathrm{Ext}(X)$ is a group and identify the inverse for X a subset of the plane. We construct a homomorphism of $\mathrm{Ext}(X)$ into $\mathrm{Hom}(\pi^1(X),\mathbb{Z})$ (where $\pi^1(X)$ is the first cohomotopy group of X), and show that this map is a bijection for X a subset of the plane.

This last theorem, which may be viewed as the principal result of the paper, has the following implications:

(1) for any essentially normal operators S and T, S is unitarily equivalent to a compact perturbation of T if and only if S and T have the same essential spectrum Λ, and $\mathrm{index}(S-\lambda I) = \mathrm{index}(T-\lambda I)$ for all λ not in Λ;

(2) an essentially normal operator T is a compact perturbation of a normal operator if and only if $\operatorname{index}(T-\lambda I) = 0$ for all λ not in the essential spectrum of T; and

(3) the family

$$\mathcal{N}+\mathcal{K} = \{N+K \mid N \text{ normal, } K \text{ compact}\}$$

is norm-closed.

The results presented in this paper are only a part of those which we have obtained on extensions. In [9] it is announced that extensions define a generalized homology functor on compact metrizable spaces which realizes K-homology [3]. Although certain basic results are given in full generality, we are mainly concerned in this paper with the implications of our results for operator theory.

This work began in the Spring of 1970 with the discovery of Theorem 3.1 (cf. [13], p. 78). Publication was delayed through the intervening period for several reasons, not the least of which was the search for the more definitive results which we now present.

§2. Sequences of Operators

In the same way that single operators correspond to extensions by $C(X)$ with X a subset of the plane (i.e. $C(X)$ singly-generated), certain sequences of operators correspond to extensions by $C(X)$ with X a compact metric space. In this section we prove two technical lemmas on such sequences of operators.

The following remarks will be needed in the first lemma. First, by the <u>joint spectrum</u> of a sequence $\{A_n\}$ of elements of a commutative C*-algebra $\mathcal{C}$, we understand the set of sequences $\{\chi(A_n)\}$, where χ is a complex *-homomorphism of $\mathcal{C}$. Next, if $(T_1, T_2, \ldots, T_n)$ is a n-tuple of mutually commuting normal operators on $\mathcal{H}$, $(\lambda_1, \lambda_2, \ldots, \lambda_n)$ is in their joint spectrum, and $\varepsilon > 0$, then there exists a unit vector ϕ such that

$$||T_k\phi - \lambda_k\phi|| < \varepsilon, \ 1 \leq k \leq n .$$

For by the spectral theorem there is a projection-valued measure E on $\mathbb{C}^n$ such that

$$T_k = \int_{\mathbb{C}^n} z_k dE(z) \ , \ 1 \leq k \leq n \ ;$$

if $U = \{z \in \mathbb{C}^n | |z_k-\lambda_k| < \varepsilon \ , \ 1 \leq k \leq n\}$ and ϕ is a unit vector in $E(U)H$, then

$$||T_k\phi - \lambda_k\phi||^2 = \int |z_k-\lambda_k|^2 d(E(z)\phi,\phi)$$

$$= \int_U |z_k-\lambda_k|^2 d(E(z)\phi,\phi) < \varepsilon^2 .$$

The point to be checked here is $E(U) \neq 0$, and this is clear, for otherwise the functions

$$f_k(z) = (\bar{z}_k-\bar{\lambda}_k)/\sum_{i=1}^{n} |z_i-\lambda_i|^2 ,$$

bounded and continuous outside U, would produce operators $S_k = \int f_k dE$ (in the C*-algebra generated by the T_k) that satisfy the relation

$$\sum_{k=1}^{n} S_k(T_k-\lambda_k I) = I ,$$

contradicting the existence of the postulated *-homomorphism.

Finally, we need a variant of Calkin's construction [10] of a faithful representation of A. Let L'' be the linear space of all sequences ξ in H^ω such that $\xi_n \to 0$ weakly. Calkin defines the inner product $(\xi,\eta)''$ as a generalized limit of (ξ_n,η_n), but instead we select a universal net $\{n(\alpha)|\alpha \in A\}$ of positive integers (that is $n(\alpha) \to \infty$, and for every set S of positive integers, either $n(\alpha)$ is eventually in S, or $n(\alpha)$ is eventually outside S; for example, let the directed set A be a free ultrafilter on the positive integers and select $n(\alpha)$ in α for each α in A) and define

$$(\xi,\eta)'' = \lim_{\alpha}(\xi_{n(\alpha)},\eta_{n(\alpha)}) .$$

This limit exists because ξ and η are bounded sequences. In fact, if $\{x_n\}$ is any sequence in a compact Hausdorff space X, then $\{x_{n(\alpha)}\}$ is a universal net in X, hence converges to each of its cluster points, and therefore converges to a limit. The rest of the construction is as in Calkin:

$$N = \{\xi \mid (\xi,\xi)'' = 0\}$$

is a subspace; the inner product defined above induces a non-degenerate inner product $(,)'$ in $L' = L''/N$; the completion L of L' is a Hilbert space; and for T in $L(H)$, the action $(T''\xi)_n = T\xi_n$ on L'' determines a bounded operator $\tilde{\rho}(T)$ on L. The map $\tilde{\rho}$ is a *-homomorphism with kernel a two-sided ideal containing K. Hence $\ker \tilde{\rho} = K$, and $\tilde{\rho}$ induces a faithful representation ρ of A on L.

<u>2.1 Lemma</u>. Let $T_1,T_2,\ldots$ be operators on H such that $T_i^*T_j - T_jT_i^*$ is compact for all i and j, and let $\{\lambda_k\}$ be in the joint spectrum of $\{\pi(T_k)\}$. Then there exists an orthonormal sequence $\{\phi_n\}$ such that

$$\lim_{n\to\infty} ||T_k\phi_n - \lambda_k\phi_n|| = 0 , \ k \geq 1 .$$

<u>Proof</u>. (Note that the $\pi(T_k)$ are mutually commuting normal elements of A by hypothesis and the Putnam-Fuglede Theorem.) Assume that pairwise orthogonal unit vectors $\phi_1,\ldots,\phi_{m-1}$ have been constructed so that

(1) $\quad ||T_k\phi_n - \lambda_k\phi_n|| < \frac{1}{n} , \ k \leq n \leq m-1 .$

With ρ as above, $\rho\pi(T_1),\dots,\rho\pi(T_m)$ are mutually commuting normal operators with joint spectrum containing $(\lambda_1,\dots,\lambda_m)$, and hence there exists a unit vector ξ' in L such that

$$||\rho\pi(T_k)\xi' - \lambda_k\xi'|| < \frac{1}{m}, \ 1 \leq k \leq m .$$

It follows that there is ξ in L'' such that

$$\text{(2)} \qquad \lim_{\alpha} ||\xi_{n(\alpha)}|| = 1$$

$$\text{(3)} \qquad \lim_{\alpha} ||T_k\xi_{n(\alpha)} - \lambda_k\xi_{n(\alpha)}|| < \frac{1}{m}, \ 1 \leq k \leq m .$$

Let Δ be one-half the minimum of the differences between the sides of the inequalities (3), and in the cofinal set of α for which

$$||T_k\xi_{n(\alpha)} - \lambda_k\xi_{n(\alpha)}|| < \frac{1}{m} - \Delta, \ 1 \leq k \leq m$$

choose α large enough that for $\psi = \xi_{n(\alpha)}$

$$\text{(4)} \qquad ||\psi|| > 1-\delta$$

$$\text{(5)} \qquad |(\psi,\phi_k)| < \delta/\sqrt{m-1}, \ 1 \leq k \leq m-1 ,$$

where $0 < \delta < \frac{1}{2}$. If

$$\phi = \psi - \sum_{k=1}^{m-1} (\psi,\phi_k)\phi_k ,$$

the unit vector $\phi_m = \phi/||\phi||$ is orthogonal to $\phi_1,\dots,\phi_{m-1}$, and

$$||T_k\phi_m-\lambda_k\phi_m|| \leq (||T_k\psi-\lambda_k\psi|| + 2||T_k|| \ ||\phi-\psi||)/||\phi||$$

for all k. Since $||\phi-\psi|| < \delta$ by (5) and $1/||\phi|| < 1 + 2\delta$ by (4), it follows that

$$||T_k\phi_m-\lambda_k\phi_m|| < \frac{1}{m} - \Delta + M\delta, \ 1 \leq k \leq m$$

where M depends only on m and the norms $||T_k||$, $k \leq m$. Thus with $\delta \leq \Delta/M$ the inequalities (1) are satisfied for $k \leq n \leq m$, and the lemma follows.

2.2 Lemma. Let $T_1, T_2, \ldots$ be operators on H such that $T_i^* T_j - T_j T_i^*$ is compact for all i and j, and let $\lambda^{(r)} = \{\lambda_k^{(r)}\}$ be in the joint spectrum of $\{\pi(T_k)\}$, $r \geq 1$. Then there exists an orthonormal sequence $\{\psi_r\}$ such that for $k \geq 1$

$$T_k = (D_k \oplus R_k) + L_k$$

relative to $[\psi_r] \oplus [\psi_r]^{\perp}$, where L_k is compact, D_k is diagonal with eigenvectors $\{\psi_r\}$ and corresponding eigenvalues $\{\lambda_k^{(r)} | r \geq 1\}$, and $\pi(R_k)$ has the same spectrum as $\pi(T_k)$.

Proof. Consider first the case of self-adjoint operators T_k. We construct recursively an orthonormal sequence $\{\psi_r\}$ such that

$$||T_k \psi_r - \lambda_k^{(r)} \psi_r|| < 1/2^r, \quad k \leq r.$$

Suppose that pairwise orthogonal unit vectors $\psi_1, \ldots, \psi_{s-1}$ have been found satisfying these inequalities for $k \leq r < s$. By Lemma 2.1, there is an orthonormal sequence $\{\phi_n\}$ such that

$$\lim_{n\to\infty} ||T_k \phi_n - \lambda_k^{(s)} \phi_n|| = 0, \quad k \geq 1.$$

Arguing as in the proof of that lemma, let $0 < \delta < \frac{1}{2}$ and choose n sufficiently large that $\phi = \phi_n$ satisfies

$$|(\phi, \psi_r)| < \delta/\sqrt{s-1}, \quad r < s$$

$$||T_k \phi - \lambda_k^{(s)} \phi|| < \delta, \quad k \leq s.$$

If $\psi = \phi - \sum_{r=1}^{s-1} (\phi, \psi_r)\psi_r$, then as before $\psi_s = \psi/||\psi||$ is a unit vector orthogonal to $\psi_1, \ldots, \psi_{s-1}$ which satisfies inequalities

$$||T_k\psi_s - \lambda_k^{(s)}\psi_s|| < M\delta \ , \ k \leq s$$

with M depending only on the norms $||T_k||$, $k \leq s$. Hence the required inequality is satisfied for $r = s$ when δ is sufficiently small.

Now if

$$T_k = \begin{bmatrix} X_k & Y_k^* \\ Y_k & R_k \end{bmatrix}$$

relative to $[\psi_r] \oplus [\psi_r]^\perp$, then

$$T_k - D_k \oplus R_k = \begin{bmatrix} X_k - D_k & Y_k^* \\ Y_k & 0 \end{bmatrix} \quad \text{and}$$

$$||(X_k-D_k)\psi_r||^2 + ||Y_k\psi_r||^2 = ||(T_k-D_k)\psi_r||^2$$

$$= ||T_k\psi_r - \lambda_k^{(r)}\psi_r||^2 < (1/2^r)^2$$

so that X_k-D_k and Y_k are Hilbert-Schmidt and $T_k - D_k \oplus R_k$ is compact. To ensure that $\pi(R_k)$ and $\pi(T_k)$ have the same spectrum, we need only use each $\lambda^{(r)}$ twice in succession in the above construction, in which case each R_k is itself a compact perturbation of an operator of the form $D_k \oplus S_k$.

The general case is reduced to the self-adjoint case as follows. As remarked before, the hypothesis and the Putnam-Fuglede Theorem imply that the $\pi(T_k)$ are mutually commuting normal elements of A . Straightforward calculation then shows that the self-adjoint sequence $\{\text{Re } T_k \ , \ \text{Im } T_k\}$ also satisfies the hypothesis of the lemma.

Results analogous to the next corollary have recently been obtained by Anderson [1], Pearcy and Salinas [30], and by Stampfli [33].

<u>2.3 Corollary</u>. If T is essentially normal and N is normal with essential spectrum contained in that of T (or N is normal on a finite-dimensional space), then $T \oplus N$ is unitarily equivalent to a compact perturbation of T.

<u>Proof</u>. Let the sequence $\{\lambda^{(r)}\}$ be dense in the essential spectrum of T, isolated points being repeated infinitely often. By Lemma 2.2

$$T = (D \oplus R) + L$$

where D is diagonal with eigenvalue sequence $\{\lambda^{(r)}\}$ and L is compact. Then D and T have the same essential spectrum, and therefore so do D and $D \oplus N$. By the Weyl-von Neumann-Berg Theorem $D \oplus N$ is unitarily equivalent to a compact perturbation of D, and hence $T \oplus N$ is unitarily equivalent to a compact perturbation of T.

§3. Unitary Elements

In the introduction it was shown that essentially normal operators T with essential spectrum lying on a simple arc are compact perturbations of normal operators, and it follows from the Weyl-von Neumann-Berg Theorem that such operators are classified up to unitary equivalence modulo the compacts by their essential spectra. In this section we suppose that the essential spectrum lies on a simple closed curve, or equivalently, the unit circle. The latter are the operators T for which $\pi(T)$ is unitary. If the essential spectrum is not the entire circle, we are of course back in the case covered in Example 1.3. In the case of the entire circle, we now show that there are countably many classes, distinguished by the integer $n = \text{ind}\, T$. As "canonical operators" we may therefore take the powers of the unilateral shift and its adjoint.

Before proceeding we recall some of the pertinent facts from index theory (cf. [17]). An operator T is said to be Fredholm if its range $R(T)$ is closed and both the null space $N(T)$ and $R(T)^{\perp}$ are finite-dimensional. The index is defined for Fredholm operators by $\text{ind}(T) = \dim N(T) - \dim R(T)^{\perp}$ and satisfies: (1) if T is Fredholm and K is compact, then $T+K$ is Fredholm and $\text{ind}(T) = \text{ind}(T+K)$; (2) if S and T are Fredholm, then ST is Fredholm and $\text{ind}(ST) = \text{ind}(S) + \text{ind}(T)$; and (3) the collection of Fredholm operators is an open subset of $L(H)$ on which the index is a norm-continuous function. Lastly, a fundamental result due to Atkinson [4] states that an operator T in $L(H)$ is Fredholm if and only if $\pi(T)$ is invertible in A and thus the set of λ for which $T-\lambda$ is not Fredholm is precisely the spectrum of $\pi(T)$ (the essential spectrum of T, c.f. [21]).

We can now classify the unitary elements of A.

3.1 Theorem. If $\pi(T)$ is unitary, then T is a compact perturbation of a unitary operator, a shift of multiplicity n, or the adjoint of a shift of multiplicity n, according as $\text{ind}\, T = 0$, $\text{ind}\, T = -n < 0$, or $\text{ind}\, T = n > 0$.

Proof. Let $\pi(T)$ be unitary. Then $T^*T - I$ is compact, and on multiplying by the inverse of $(T^*T)^{1/2} + I$, we find that $(T^*T)^{1/2} - I$ is compact. If $T = W(T^*T)^{1/2}$ is the polar decomposition, it follows that $T = W + K$ with K compact. We can assume that $\text{ind}\, T \leq 0$, by taking adjoints if necessary. Then $\text{ind}\, W \leq 0$, $\dim N(W) \leq \dim R(W)^{\perp}$, there exists a partial isometry L with initial space $N(W)$ and final space contained in $R(W)^{\perp}$, and therefore T is a compact perturbation of the isometry $V = W + L$.

Now by the "Wold decomposition" $V = U \oplus S$ with U unitary and S a unilateral shift of suitable multiplicity (either summand may be absent); since

$$\text{ind}\, T = \text{ind}\, V = \text{ind}\, U + \text{ind}\, S = \text{ind}\, S\ ,$$

that multiplicity is $-\text{ind } T$. If $\text{ind } T = 0$, the shift summand is absent and T is a compact perturbation of the unitary U . If $\text{ind } T < 0$, it follows from Corollary 2.3 that the shift summand "absorbs" U , and consequently T is unitarily equivalent to a compact perturbation of S .

It is of interest to work out explicitly for this case some of the reductions of the introduction. If $(\mathcal{E},\phi)$ is an extension of $\mathcal{K}$ by $C(S^1)$, where S^1 is the unit circle, then the function $\chi(z) = z$ can be pulled back to one of the canonical operators of Theorem 3.1. This enables us to obtain canonical realizations for these extensions. Take $\mathcal{H}$ to be the Hardy space H^2 on the circle, and for f in L^∞ let T_f be the corresponding Toeplitz operator defined on H^2 by $T_f h = P(fh)$, where P is the orthogonal projection of L^2 on H^2 . For each nonzero integer n , let

$$\mathcal{E}_n = \{T_{f\circ\chi^n} + K \mid f \in C(S^1) , K \in \mathcal{K}\}$$

and define ϕ_n on $\mathcal{E}_n$ by

$$\phi_n(T_{f\circ\chi^n} + K) = f .$$

In addition, for $f \in C(S^1)$ let M_f be multiplication by f in L^2 , let

$$\mathcal{E}_0 = \{M_f + K \mid f \in C(S^1) , K \in \mathcal{K}\} ,$$

and define ϕ_0 on $\mathcal{E}_0$ by $\phi_0(M_f + K) = f$. The relations

$$cT_f = T_{cf}$$

$$T_f + T_g = T_{f+g}$$

$$(T_f)^* = T_{\bar{f}}$$

$$T_f T_g = T_{fg} + K \quad [12]$$

$$||f|| = ||T_f|| \leq ||T_f + K|| \quad [7;11]$$

imply that $\mathcal{E}_n$ is a C*-algebra and that ϕ_n is a well-defined *-homomorphism with kernel $\mathcal{K}$, so that $(\mathcal{E}_n, \phi_n)$ is an extension of $\mathcal{K}$ by $C(S^1)$. The same is readily seen to be true of $(\mathcal{E}_0, \phi_0)$. Since

$$\phi_n(T_{\chi^n}) = \phi_0(M_\chi) = \chi ,$$

the "generators" of these extensions are respectively the unilateral shift of multiplicity n, its adjoint, and M_χ. Therefore these extensions form a complete system of representatives.

It is also interesting to note that there is at hand a natural system of representatives for operators with essential spectrum a simple closed curve Γ.

<u>3.2 Corollary</u>. Let η be an orientation-preserving homeomorphism of S^1 onto the simple closed curve Γ. If T is an essentially normal operator on $\mathcal{H}$ with essential spectrum Γ, and $\mathrm{ind}(T-\lambda I) = n$ for λ inside Γ, then T is unitarily equivalent to a compact perturbation of M_η or $T_{\eta\circ\chi^{-n}}$ according as $n = 0$ or $n \neq 0$.

<u>Proof</u>. If $(\mathcal{E}, \phi)$ is the extension of $\mathcal{K}$ by $C(\Gamma)$ determined by T, then $(\mathcal{E}, \eta^*\phi)$ is an extension of $\mathcal{K}$ by $C(S^1)$ and thus is equivalent to one of the canonical extensions $(\mathcal{E}_k, \phi_k)$ listed above. Since $\eta^*\phi(T) = \eta$, $\phi_k(T_{\eta\circ\chi^k}) = \eta$ for $k \neq 0$, and $\phi_0(M_\eta) = \eta$, it follows that T is unitarily equivalent to a compact perturbation of one of $T_{\eta\circ\chi^k}$ or M_η. Since this relation preserves the index, $\mathrm{ind}(T_{\eta\circ\chi^k} - \lambda) = -k$, and $\mathrm{ind}(M_\eta - \lambda) = 0$ [17], the proof is complete.

<u>§4. Extensions</u>

We make the standing assumption that all spaces $X, Y, \ldots$ are compact and metrizable. By an extension of $\mathcal{K}$ by $C(X)$ one usually means a short exact sequence

$$0 \to K \underset{\rho}{\to} E \underset{\phi}{\to} C(X) \to 0$$

of C*-algebras and *-homomorphisms; another such sequence

$$0 \to K \underset{\rho'}{\to} E' \underset{\phi'}{\to} C(X) \to 0$$

is equivalent to the first if there is a *-isomorphism $\psi : E' \to E$ such that the following diagram commutes:

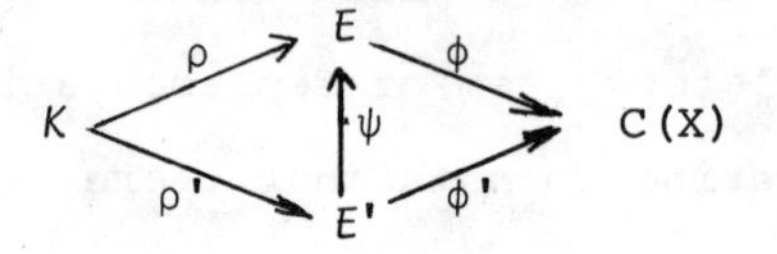

We make several modifications. Assume that E is represented as a concrete C*-algebra of operators acting on a separable Hilbert space H. Then ρ defines a representation of K on H, and is thus a direct sum of irreducible representations and the zero representation. (c.f.[17], Thm. 5.40).

This means that there exist a decomposition $H = \Sigma \oplus H_i \oplus N$ and unitaries $U_i : H \to H_i$ such that $\rho(K) = \Sigma \oplus U_i K U_i^* \oplus 0$ for all $K \in K$. Now $\text{im}\,\rho = \ker \phi$ is an ideal of E, and it follows that the H_i and N are invariant for E : $T = \Sigma \oplus T_i \oplus T'$ for each $T \in E$. Moreover $T\rho(K) \in \ker \phi$ so $T\rho(K) = \rho(K')$ for some $K' \in K$; thus $T_i U_i K U_i^* = U_i K' U_i^*$ for all i, so that $T'' = U_i^* T_i U_i$ is independent of i, and hence $T = \Sigma \oplus U_i T'' U_i^* \oplus T'$. If we now let

$$E' = \{T''\oplus T' \mid T \in E\};\ \rho'(K) = K\oplus 0;\ \phi'(T''\oplus T') = \phi(T)$$

it is clear that

$$0 \to K \underset{\rho'}{\to} E' \underset{\phi'}{\to} C(X) \to 0$$

is exact and equivalent to the given sequence.

Now observe that $E'|H \supset K$, that the quotient is *-isomorphic to $C(X_1)$ for some closed $X_1 \subset X$, and that $E'|N$ is *-isomorphic to $C(X_2)$ for some closed $X_2 \subset X$ such that $X_1 \cup X_2 = X$.

From this analysis we conclude that there will be no significant loss of generality in considering only those extensions for which ρ does not contain the zero representation (that is $N = \{0\}$), and that (up to equivalence) this amounts to confining ourselves to extensions for which ρ is inclusion.

Next we assume that E contains the identity operator on H. In any case, there exists a projection $P \in E$ such that $\phi(P) = 1$ [10], and the compression of E to PH produces an extension of K by $C(X)$ that does contain the identity. The compression of E to $(I-P)H$ is the compact operators on that space, which therefore may be disregarded without essential loss.

Finally, having "fixed" the compact operators by making ρ the inclusion map, the appropriate notion of equivalence is obtained by deleting the left side of the above diagram (or by allowing an automorphism of K).

4.1 Definition. Let X be compact and metrizable. An extension of K by $C(X)$ is a pair (E,ϕ), where E is a C*-subalgebra of $L(H)$ that contains $K(H)$ and I, and ϕ is a *-homomorphism of E onto $C(X)$ with kernel K. Extensions (E_1,ϕ_1) on H_1 and (E_2,ϕ_2) on H_2 are equivalent if there exists a *-isomorphism $\psi : E_1 \to E_2$ such that $\phi_1 = \phi_2\psi$. The set of equivalence classes of extensions of K by $C(X)$ is denoted by $\mathrm{Ext}(X)$.

We remark that the presence of the compact operators implies that equivalences are spatial. For $\psi(K(H_1)) = K(H_2)$ (since $\ker \phi_i = K(H_i)$), and therefore [15; 4.1.8] there exists unitary $U : H_1 \to H_2$ with $\psi(K) = UKU^*$ for all $K \in K(H_1)$. Then for $T \in E_1$ and $K \in K(H_1)$ we have

$$(UTU^*)(UKU^*) = U(TK)U^* = \psi(TK) = \psi(T)UKU^*$$

and consequently $\psi(T) = UTU^*$.

An equivalent definition of extension is possible in terms of maps from $C(X)$ into the Calkin algebra. This will be our usual working definition. We point out that in [34] Thayer studied the analogous concept in which $C(X)$ is replaced by an arbitrary C*-algebra and in this context independently obtained Theorem 3.1 and the fact that all extensions for totally disconnected spaces are trivial.

4.2 Definition. An extension of K by $C(X)$ is a *-monomorphism τ of $C(X)$ into the Calkin algebra A such that $\tau(1) = 1$. Extensions τ_1 and τ_2 are equivalent if there is a *-isomorphism $\mu : A(H_1) \to A(H_2)$ induced by a unitary $U : H_1 \to H_2$ such that $\mu\tau_1 = \tau_2$.

It is easy to see that these notions are connected by a bijection that respects equivalence (to (E,ϕ) corresponds the unique map $\tau : C(X) \to \pi(E)$ with $\tau\phi = \pi|E$, and to τ corresponds (E,ϕ) with $E = \pi^{-1}(\text{im}\,\tau)$ and $\phi : E \to C(X)$ the unique map with $\tau\phi = \pi|E)$.

This formulation suggests once again the weakened version of equivalence (raised in the Introduction) obtained by replacing the unitary $U : H_1 \to H_2$ by an arbitrary operator $T: H_1 \to H_2$ that induces a *-isomorphism $A(H_1) \to A(H_2)$. We refer to this notion as weak equivalence.

4.3 Theorem. Weakly equivalent extensions are equivalent.

Proof. Let $\tau_1 : C(X) \to A(H_1)$ and $\tau_2 : C(X) \to A(H_2)$ be weakly equivalent via $T : H_1 \to H_2$. It can be assumed that $H_1 = H_2 = H$ by replacing τ_1 and τ_2 by equivalent extensions. Then $\mu : A \to A$ is given by $\mu(a) = \pi(T)a\pi(T)^*$, and it follows that $\pi(T)$ is a unitary element of A .

It will be sufficient to show that for any extension τ and integer n there exists unitary $\pi(S)$ with index n that commutes with $\operatorname{im}\tau$. For then, with $\tau = \tau_1$ and $n = -\operatorname{ind} T$, the operators T and TS implement the same mapping on $\operatorname{im}\tau$ and TS is a compact perturbation of a unitary by Theorem 3.1.

Let $\{f_k\}$ be dense in $C(X)$, let $\pi(T_k) = \tau(f_k)$, let $\lambda = \{\lambda_k\}$ be in the joint spectrum of $\{\pi(T_k)\}$, and apply Lemma 2.2 with $\lambda^{(r)} = \lambda$ for all $r \geq 1$ to obtain an orthonormal sequence $\{\phi_r\}$ such that

$$T_k = (\lambda_k I \oplus R_k) + L_k \quad , \quad k \geq 1$$

relative to $[\phi_r] \oplus [\phi_r]^{\perp}$, with L_k compact. If S is the identity in $[\phi_r]^{\perp}$ and is isometric or coisometric of index n in $[\phi_r]$, then $[S,T_k]$ is compact and

$$\pi(S)\tau(f_k) = \pi(S)\pi(T_k) = \tau(f_k)\pi(S)$$

so that $\pi(S)$ commutes with $\operatorname{im}\tau$ by continuity.

<u>4.4 Corollary</u>. Let $S,T \in L(H)$ be essentially normal. If there exists $V \in L(H)$ such that $\pi(V)$ is unitary and $V^*SV - T$ is compact, then there exists unitary $U \in L(H)$ such that $U^*SU - T$ is compact.

Next we introduce the promised addition of extensions.

<u>4.5 Definition</u>. The <u>sum</u> of <u>extensions</u> (E_1,ϕ_1) on H_1 and (E_2,ϕ_2) on H_2 of K by $C(X)$ is the extension of K by $C(X)$ defined by

$$(E,\phi) = (E_1,\phi_1) + (E_2,\phi_2) \ ,$$

$$E = \{(T_1 \oplus T_2) + K \mid T_i \in E_i \ , \ \phi_1(T_1) = \phi_2(T_2) \ , \ K \in K(H_1 \oplus H_2)\} \ ,$$

$$\phi((T_1 \oplus T_2) + K) = \phi_1(T_1) = \phi_2(T_2) \ .$$

It is clear that (E,ϕ) is an extension of K by $C(X)$. Moreover, if (E_1,ϕ_1) is equivalent to (E_1',ϕ_1'), then $(E_1,\phi_1) + (E_2,\phi_2)$ is equivalent to $(E_1',\phi_1') + (E_2,\phi_2)$, and therefore addition of extensions induces a binary operation on $\mathrm{Ext}(X)$. This operation is associative and commutative. In the next section we show that the class of trivial extensions is the identity element, and in subsequent sections we prove the existence of inverses, so that $\mathrm{Ext}(X)$ is an abelian group.

The corresponding operation on maps $\tau_1 : C(X) \to A(H_1)$ and $\tau_2 : C(X) \to A(H_2)$ is simply

$$\tau_1 + \tau_2 : C(X) \to A(H_1 \oplus H_2)$$

defined by $(\tau_1+\tau_2)(f) = \tau_1(f) \oplus \tau_2(f)$, where $A(H_1) \oplus A(H_2)$ is regarded in the natural way as a subalgebra of $A(H_1 \oplus H_2)$.

<u>4.6 Remark</u>. In the Introduction it was pointed out that any essentially normal operator T determines an extension (E_T,ϕ_T) of K by $C(X)$, where E_T is generated by T, I and K, $\phi_T(T) = \chi$, and $X = \sigma(\pi(T))$. (Moreover, any extension for a subset of $\mathbb{C}$ arises in this way.) The corresponding map $\tau : C(X) \to A$ is simply the functional calculus for $\pi(T)$: $\tau(f) = f(\pi(T))$. Also, the sum of (E_S,ϕ_S) and (E_T,ϕ_T) is $(E_{S\oplus T},\phi_{S\oplus T})$, where $\sigma(\pi(S)) = \sigma(\pi(T))$.

We conclude this section with another useful operation on extensions. The customary notation $X_1 \vee X_2$ is employed for the disjoint union of spaces.

<u>4.7 Definition</u>. The <u>disjoint sum</u> of extensions (E_1,ϕ_1) of K by $C(X_1)$ on H_1 and (E_2,ϕ_2) of K by $C(X_2)$ on H_2 is the extension of K by $C(X_1 \vee X_2)$ defined by

$$(E,\phi) = (E_1,\phi_1) \vee (E_2,\phi_2)$$

$$E = \{(T_1 \oplus T_2) + K \mid T_i \in E_i \ , \ K \in L(H_1 \oplus H_2)\}$$

$$\phi((T_1 \oplus T_2)+K)\mid X_i = \phi_i(T_i) \ , \ i=1,2 \ .$$

The corresponding operation on maps $\tau_1 : C(X_1) \to A(H_1)$ and $\tau_2 : C(X_2) \to A(H_2)$ is

$$\tau_1 \vee \tau_2 : C(X_1 \vee X_2) \to A(H_1 \oplus H_2)$$

defined by $(\tau_1 \vee \tau_2)(f) = \tau_1(f|X_1) \oplus \tau_2(f|X_2)$.

<u>4.8 Theorem</u>. The operation $\vee$ induces an isomorphism $\lambda : \mathrm{Ext}(X_1) \oplus \mathrm{Ext}(X_2) \to \mathrm{Ext}(X_1 \vee X_2)$.

<u>Proof</u>. It is easily shown that the class of $\tau_1 \vee \tau_2$ depends only on those of τ_1 and τ_2, and that the induced map is a homomorphism. The inverse map $\mu : \mathrm{Ext}(X_1 \vee X_2) \to \mathrm{Ext}(X_1) \oplus \mathrm{Ext}(X_2)$ is defined as follows. Let $\tau : C(X_1 \vee X_2) \to A(H)$ be an extension, let χ_1 be the characteristic function of X_1 , let p be the projection $\tau(\chi_1)$, and define $\tau_1 : C(X_1) \to pAp$ by $\tau_1 = \tau|C(X_1)$. Here we regard $C(X_1)$ as a subalgebra of $C(X_1 \vee X_2)$ by making the functions in $C(X_1)$ zero on X_2 . Since $f\chi_1 = f$ for all $f \in C(X_1)$, $\mathrm{im}\tau_1 \subset pAp$. If P is any projection with $\pi(P) = p$, then pAp is naturally *-isomorphic to $A(PH)$, so we can view τ_1 as an extension by regarding it as a map into $A(PH)$. The extension $\tau_2 : C(X_2) \to A((I-P)H)$ is obtained similarly, and we take μ to be the map induced by $\tau \to (\tau_1,\tau_2)$. The construction makes it clear that $\lambda\mu = \mathrm{id}$. To obtain $\mu\lambda = \mathrm{id}$ it is necessary to show that the

definition of μ is independent of the choice of the projection P. Let Q be any other projection with $\pi(Q) = p$ and let $\tau_1' : C(X_1) \to A(QH)$ be the associated extension. To see that τ_1 and τ_1' are equivalent, choose isometries $U, V \in L(H)$ with $UH = PH$ and $VH = QH$, and let $\alpha : A(PH) \to A(H)$, $\beta : A(QH) \to A(H)$ be the corresponding *-isomorphisms. It is enough to show $\alpha\tau_1$ equivalent to $\beta\tau_1'$. If $f \in C(X_1)$ and $\pi(T) = \tau(f)$, then $\alpha\tau_1(f) = \pi(U^*TU)$ and $\beta\tau_1'(f) = \pi(V^*TV)$, so

$$\pi(V^*U)\alpha\tau_1(f) = \beta\tau_1'(f)\pi(V^*U) .$$

But $\pi(V^*U)$ is unitary, so $\alpha\tau_1$ and $\beta\tau_1'$ are weakly equivalent, hence equivalent by Theorem 4.3. Returning to $\mu\lambda = \mathrm{id}$, we now take P to be the projection of $H_1 \oplus H_2$ on H_1 to get $\mu[\tau_1 \vee \tau_2] = ([\tau_1],[\tau_2])$.

<u>4.9 Remark</u>. This proof suggests that one associate to a pair of projections P, Q with $\pi(P) = \pi(Q)$ the integer $\mathrm{ind}(V^*U)$, where U, V are isometries with $UH = PH$ and $VH = QH$. This might be called the <u>essential codimension</u> of QH in PH, since it is the usual codimension when $Q \leq P$. The essential codimension is zero if and only if there is unitary W with $W^*PW = Q$ and $\pi(W) = 1$. This will always be the case if $||P-Q|| < 1$.

§5. Trivial Extensions

<u>5.1 Definition</u>. The extension (E,ϕ) of K by $C(X)$ is <u>trivial</u> (or split) if there exists a *-monomorphism σ of $C(X)$ into E such that $\sigma(1) = I$ and $\phi\sigma$ is the identity on $C(X)$. Equivalently, $\tau : C(X) \to A(H)$ is trivial if there exists a *-monomorphism $\sigma : C(X) \to L(H)$ such that $\sigma(1) = I$ and $\pi\sigma = \tau$.

5.2 Remark. The requirement that $\sigma(1) = I$ is not essential. In any case $\sigma(1)$ is a projection P ; if $x_0 \in X$, and σ' is defined by $\sigma'(f) = \sigma(f) + f(x_0)(I-P)$, then σ' is a *-monomorphism with $\sigma'(1) = I$ and $\pi\sigma' = \tau$.

5.3 Theorem. If X is a compact metric space, then there exists a trivial extension of K by $C(X)$, and any two trivial extensions are equivalent.

Proof. Let $\{x_n\}$ be dense in X , isolated points being repeated infinitely often. Take $H = \ell^2$ and define $\tau : C(X) \to A$ by

$$\tau(f) = \pi \ \mathrm{diag}(f(x_n)) \ ,$$

where $\mathrm{diag}(f(x_n))$ is diagonal in the usual orthonormal basis $\{e_n\}$ for ℓ^2 . This is obviously a *-homomorphism that factors through π , and if $\tau(f) = 0$, then $\mathrm{diag}(f(x_n))$ is compact, $f(x_n) \to 0$, and $f = 0$ since $\{x_k, x_{k+1}, \ldots\}$ is dense in X for all k .

We show next that any two extensions of this type are equivalent. Let τ' arise from another such sequence $\{y_n\}$ and choose a permutation α of the positive integers such that $d(x_n, y_{\alpha(n)}) \to 0$, d being the metric on X . (The existence of such permutations is proved in [22]. For a more elementary proof, assume that sets $\{n_1, \ldots, n_{2k}\}$ and $\{m_1, \ldots, m_{2k}\}$ of distinct positive integers have been chosen, each containing $1, 2, \ldots, k$, such that $d(x_{n_i}, y_{m_i}) < 1/i$. It is clear how to extend to $2k+2$, and iteration leads to permutations β, γ such that $d(x_{\beta(n)}, y_{\gamma(n)}) \to 0$, so we may take $\alpha = \gamma\beta^{-1}$.) If U is defined on ℓ^2 by $Ue_n = e_{\alpha(n)}$, we have

$$U \ \mathrm{diag}(f(x_n))e_n = f(x_n)e_{\alpha(n)}$$

$$\mathrm{diag}(f(y_n))Ue_n = f(y_{\alpha(n)})e_{\alpha(n)}$$

so that $U\operatorname{diag}(f(x_n)) - \operatorname{diag}(f(y_n))U$ is compact. Hence $\pi(U)\tau(f) = \tau'(f)\pi(U)$ for all f in $C(X)$ as required.

The proof is completed by showing that any trivial extension arises in the above manner. Let $\tau : C(X) \to A$ be trivial, so that $\tau = \pi\sigma$ for a *-monomorphism $\sigma : C(X) \to L(H)$. By the spectral theorem there exists a projection-valued measure E on the Borel subsets of X such that $\sigma(f) = \int f dE$ for all f in $C(X)$. Let $\{U_n\}$ be a basis of open sets for X and let Z be the C*-algebra generated by the projections $E_n = E(U_n)$. Then Z contains $\sigma(C(X))$ and has a single self-adjoint generator H ([32], p. 293 ; the operator $H = \sum_n 3^{-n}(2E_n - I)$ will do). If Λ is the spectrum of H and $\rho : C(\Lambda) \to Z$ is the *-isomorphism $\rho(g) = g(H)$, then the embedding $\rho^{-1}\sigma$ of $C(X)$ in $C(\Lambda)$ is induced by a surjection $p : \Lambda \to X$, so that

$$\sigma(f) = \rho(f \circ p) = (f \circ p)(H) \text{ , for } f \text{ in } C(X) \text{ .}$$

By Weyl's Theorem $H = D + K$ with D diagonal and K compact. The spectra of H and D have the same limit points, so that by making a compact perturbation of D it may be assumed that D has spectrum Λ. It follows that

$$\sigma(f) = (f \circ p)(D) + K_f \text{ , } K_f \text{ compact}$$

for all f in $C(X)$, since $\pi(f \circ p)(D) = (f \circ p)(\pi(D))$ and similarly for H. If $D = \operatorname{diag}(\lambda_n)$, then $\{\lambda_n\}$ is dense in Λ, and consequently $\{x_n\} = \{p(\lambda_n)\}$ is dense in X. Moreover since $(f \circ p)(D) = \operatorname{diag}(f(x_n))$ we have

$$\tau(f) = \pi\sigma(f) = \pi((f \circ p)(D)) = \pi \operatorname{diag}(f(x_n))$$

for all f in $C(X)$, so that τ arises as above from $\{x_n\}$. (The requirement that $x = x_n$ for infinitely many n when x is isolated is automatic, for otherwise we would have $\tau(f) = 0$ for f the characteristic function of $\{x\}$.)

This argument is modeled on Halmos' elegant proof [23] of Berg's generalization of the Weyl-von Neumann Theorem (but avoids the use of cross sections). Indeed, it contains the following extension of that result:

5.4 Corollary. If $\mathcal{Z}$ is a separable commutative C*-subalgebra of $\mathcal{L}(\mathcal{H})$, then there exists an orthonormal basis $\{\phi_n\}$ of $\mathcal{H}$ such that each operator in $\mathcal{Z}$ is a compact perturbation of an operator that is diagonal relative to $\{\phi_n\}$. In particular, the same conclusion holds for any countable family of mutually commuting normal operators on $\mathcal{H}$.

Next we show that the trivial class functions as the identity in $\mathrm{Ext}(X)$.

5.5 Theorem. Let $(\mathcal{E}_0, \phi_0)$ be a trivial extension of $\mathcal{K}$ by $C(X)$, X compact metric. Then for any extension $(\mathcal{E}, \phi)$ of $\mathcal{K}$ by $C(X)$, $(\mathcal{E}, \phi) + (\mathcal{E}_0, \phi_0)$ is equivalent to $(\mathcal{E}, \phi)$.

Proof. Let $\{f_k\}$ be dense in $C(X)$, let $\{x_r\}$ be dense in X, isolated points being repeated infinitely often, let $\lambda_k^{(r)} = f_k(x_r)$, and let T_k be in $\mathcal{E}$ with $\phi(T_k) = f_k$. Then $T_i^* T_j - T_j T_i^*$ is compact for all i and j, and $\lambda^{(r)} = \{\lambda_k^{(r)} \mid k \geq 1\}$ is in the essential joint spectrum of $\{T_k\}$, so the hypotheses of Lemma 2.2 are satisfied. Let $\{\psi_r\}$ be as provided by that lemma. Now in the decomposition

$$T = \begin{bmatrix} S & L \\ M & R \end{bmatrix}, \quad T \text{ in } \mathcal{E}$$

relative to $[\psi_r] \oplus [\psi_r]^{\perp}$, the off-diagonal entries L and M are compact, because this is true for the dense set $T_k + K$, K compact. Let

$$\mathcal{E}_1 = \{S|T \in \mathcal{E}\} \ , \ \mathcal{E}_2 = \{R|T \in \mathcal{E}\}$$

and define $\phi_i : \mathcal{E}_i \to C(X)$ by

$$\phi_1(S) = \phi_2(R) = \phi(T) \ .$$

Then the $(\mathcal{E}_i, \phi_i)$ are extensions of $\mathcal{K}$ by $C(X)$ with sum $(\mathcal{E}, \phi)$. The only non-obvious part of this assertion is that the ϕ_i are well-defined. If T is in $\mathcal{E}$ with $S = 0$, $f = \phi(T)$, and $f_{k_j} \to f$, then

$$||\pi(S_{k_j} \oplus R_{k_j}) - \pi(0 \oplus R)|| = ||\pi(T_{k_j} - T)||$$
$$= ||f_{k_j} - f|| \to 0$$

so $\pi(S_{k_j}) \to 0$. But

$$\pi(S_k) = \pi \ \mathrm{diag}(f_k(x_1), f_k(x_2), \ldots,)$$

and it follows that $f_{k_j} \to 0$ since $\{x_n, x_{n+1}, \ldots\}$ is dense in X for all n. Hence $\phi_1(S) = f = 0$. The same argument works for ϕ_2 because R_k has the same essential spectrum as T_k so that

$$||\pi(R_k)|| = ||\pi(T_k)|| = ||f_k|| \ .$$

To complete the proof we need only show that $(\mathcal{E}_1, \phi_1)$ is trivial, for then $(\mathcal{E}_0, \phi_0) + (\mathcal{E}_1, \phi_1)$ will be trivial, hence equivalent to $(\mathcal{E}_1, \phi_1)$ by Theorem 5.3, and consequently

$$(\mathcal{E}_0, \phi_0) + (\mathcal{E}, \phi) = (\mathcal{E}_0, \phi_0) + (\mathcal{E}_1, \phi_1) + (\mathcal{E}_2, \phi_2)$$

will be equivalent to $(\mathcal{E}, \phi)$.

To show that $(\mathcal{E}_1, \phi_1)$ is trivial we define the map σ from $C(X)$ to $\mathcal{L}([\psi_r])$ by $\sigma(f) = \mathrm{diag}(f(x_1), f(x_2), \ldots.)$. This map is a *-isomorphism and $\phi_1 \sigma = \mathrm{id}$ since they agree on the dense set $\{f_k\}$:

$\phi_1\sigma(f_k) = \phi_1(S_k) = \phi(T_k) = f_k$. Thus (E_1,ϕ_1) is trivial and the proof is complete.

The existence of an identity element in Ext(X) allows us to assign to each continuous map $p : X \to Y$ a homomorphism $p_* : \mathrm{Ext}(X) \to \mathrm{Ext}(Y)$ in such a way that Ext is a covariant functor. Indeed, let $p^* : C(Y) \to C(X)$ be the dual *-homomorphism; then for any extension $\tau : C(X) \to A(H_1)$, define $p_*\tau : C(Y) \to A(H_1 \oplus H_2)$ by $p_*\tau = \tau \circ p^* + \tau_Y$, where τ_Y is the trivial extension for Y . (The summand τ_Y is included to make $p_*\tau$ injective; if p is surjective then $\tau \circ p^*$ is already injective and $p_*\tau$ is equivalent to $\tau \circ p^*$ by Theorem 5.5.) It is clear that $(\mathrm{id}_X)_* = \mathrm{id}_{\mathrm{Ext}(X)}$, and that if $q : Y \to Z$, then $(qp)_* = q_* p_*$.

<u>5.6 Remark</u>. If the extension (E_T,ϕ_T) determined by an essentially normal operator T is trivial, then T is a compact perturbation of a normal operator ($T - \sigma_0(\chi)$ is in $\ker \phi_T = K$, where σ_0 is the trivializing map). Conversely, if N is normal then (E_N,ϕ_N) is trivial (let N' be the compact perturbation of N obtained by moving the isolated eigenvalues of finite multiplicity into $\sigma(\pi(N))$, so that $\sigma(N') = \sigma(\pi(N))$, and let $\sigma_0(f) = f(N'))$. In particular, the uniqueness of the trivial extension is, for subsets of $\mathbb{C}$, simply the Weyl-von Neumann-Berg Theorem. Also, the fact that the trivial class is the identity element reduces, for subsets of $\mathbb{C}$, to the "absorption lemma" 2.3.

For the remainder of this section we consider several possible alternative descriptions of triviality. To begin with, if $\tau : C(X) \to A(H)$ is trivial with trivializing map $\sigma : C(X) \to L(H)$, then for any invertible $f \in C(X)$, $\tau(f)$ is invertible so $\mathrm{ind}\,\tau(f)$ is meaningful by Atkinson's Theorem, and in fact

$$\mathrm{ind}\,\tau(f) = \mathrm{ind}\,\sigma(f) = 0$$

since $\sigma(f)$ is a normal operator. We ask whether this requirement characterizes triviality. If $\pi^1(X)$ is the first cohomotopy group of X (the group of homotopy classes of continuous maps $X \to \mathbb{C}^* = \mathbb{C}\setminus\{0\}$), this suggests that one investigate the map

$$\gamma_X : \text{Ext } X \to \text{Hom}(\pi^1(X), \mathbb{Z})$$

defined by $\gamma[\tau][f] = \text{ind}\tau(f)$, where $\tau : C(X) \to A$ is any extension and $f : X \to \mathbb{C}^*$ any continuous function. It is clear that $\text{ind } \tau(f)$ depends only on the equivalence class $[\tau]$ of τ and the homotopy class $[f]$ of f . Moreover

$$\text{ind } \tau(fg) = \text{ind } \tau(f)\tau(g) = \text{ind } \tau(f) + \text{ind } \tau(g) ,$$

so $\gamma[\tau] : \pi^1(X) \to \mathbb{Z}$ is a homomorphism; and since

$$\text{ind}(\tau_1+\tau_2)(f) = \text{ind}(\tau_1(f)\oplus\tau_2(f)) = \text{ind } \tau_1(f) + \text{ind } \tau_2(f) ,$$

γ is a homomorphism.

Our question now becomes: is γ_X injective? This is true for X a subset of $\mathbb{C}$ (§10), but is false in general [9].

<u>5.7 Theorem</u>. Im γ_X is a subgroup; consequently Ext X is a group whenever γ_X is injective.

<u>Proof</u>. Let $T \to T^t$ be the operation of transposing in $L(H)$, relative to a fixed orthonormal basis of H , and for any extension (E,ϕ) let (E^t,ϕ^t) be the extension with $\phi^t(T^t) = \phi(T)$. Since $\text{ind } T^t = -\text{ind } T$ we have $\gamma(E,\phi) + \gamma(E^t,\phi^t) = 0$ and the result follows.

In a different direction, it was demonstrated in the proof of Theorem 5.3 that for any trivial τ , im τ is contained in a commutative C*-subalgebra of A generated by projections (equivalently, one having a totally disconnected maximal ideal space). We now demonstrate the converse. Using this it can be proved directly that all extensions are trivial for certain X (this is the case for the closure of the graph of $\sin(1/t)$, $0 < t \leq 1$; we omit the details).

5.8 Theorem. If $\tau : C(X) \to \mathcal{A}$ is an extension with image contained in a commutative C*-subalgebra of $\mathcal{A}$ generated by projections, then τ is trivial.

Proof. We can assume that $\operatorname{im} \tau$ is contained in a C*-subalgebra $\mathcal{D}$ generated by a countable commutative family of projections. Then, as in the proof of Theorem 5.3, $\mathcal{D}$ is *-isomorphic to $C(\Lambda)$ for some $\Lambda \subset \mathbb{R}$. If $\tau_0 : C(\Lambda) \to \mathcal{D}$ is a *-isomorphism, the composition

$$C(X) \underset{\tau}{\to} \mathcal{D} \underset{\tau_0^{-1}}{\to} C(\Lambda)$$

is injective and thus is dual to a surjection $p : \Lambda \to X$; that is, $\tau_0^{-1}\tau = p^*$ or $\tau = \tau_0 p^* = p_*(\tau_0)$. But τ_0 is trivial (by the Weyl-von Neumann Theorem, since $\Lambda \subset \mathbb{R}$) and therefore so is τ.

5.9 Remark. This proof contains an idea that will be useful many times: a separable, commutative C*-subalgebra $\mathcal{B}$ of $\mathcal{A}$ that contains the range of an extension $\tau: C(X) \to \mathcal{A}$ gives rise to a compact metric space X', a surjection $p : X' \to X$, and an extension $\tau' : C(X') \to \mathcal{A}$, such that $p_*(\tau') = \tau$.

An important special case occurs when $\mathcal{B}$ is generated by $\operatorname{im} \tau$ and a commuting projection e. Define $\tau_1 : C(X) \to e\mathcal{A}e$ and $\tau_2 : C(X) \to (1-e)\mathcal{A}(1-e)$ by $\tau_1 = e\tau$ and $\tau_2 = (1-e)\tau$. These maps are *-homomorphisms, so their kernels are determined by closed subsets B, C of X; then the induced maps on the quotients determine elements b and c of $\operatorname{Ext}(B)$ and $\operatorname{Ext}(C)$ (strictly speaking, this entails choosing a projection E with $\pi(E) = e$, but as shown in the proof of Theorem 4.8, all choices lead to equivalent extensions). We have $X = B \cup C$, and X' is homeomorphic to $B \vee C$ in such a way that $p : X' \to X$ becomes the identification map (to see this, regard X and X' as the spaces of complex homomorphisms of $\operatorname{im} \tau$ and $\mathcal{B}$, and p as restriction). Moreover, if $a \in \operatorname{Ext}(X)$ is determined by τ and $i : B \to X$, $j : C \to X$ are the inclusion maps,

$$a = p_*(b \vee c) = i_*(b) + j_*(c) .$$

Thus we say that e "splits" a into b and c . The two expressions for a are equivalent ways of looking at this splitting. Formally, if

$$\lambda : \operatorname{Ext}(B) \oplus \operatorname{Ext}(C) \to \operatorname{Ext}(B \vee C)$$

is the isomorphism of Theorem 4.8, and

$$\beta : \operatorname{Ext}(B) \oplus \operatorname{Ext}(C) \to \operatorname{Ext}(X)$$

is defined by $\beta(b,c) = i_*(b) + j_*(c)$, then $\beta = p_*\lambda$. To say that $a \in \operatorname{Ext}(X)$ splits relative to $X = B \cup C$ is to say that $a \in \operatorname{im} \beta$, or equivalently, $a \in \operatorname{im} p_*$. Thus if $X = B \cup C$ with $B \cap C = \emptyset$, Theorem 4.8 says that every $a \in \operatorname{Ext}(X)$ splits. Later we will show, by constructing suitable projections, that if $X \subset \mathbb{C}$, $X = B \cup C$, and $B \cap C$ is a singleton or a line segment, then again every $a \in \operatorname{Ext}(X)$ splits.

<u>5.10 Corollary</u>. If $\tau : C(X) \to \mathcal{A}$ is an extension and if $\mathcal{C}$ is a commutative C*-subalgebra of $\mathcal{L}(\mathcal{H})$ such that $\pi(\mathcal{C}) \supset \operatorname{im} \tau$, then τ is trivial.

<u>Proof</u>. By the spectral theorem, $\mathcal{C}$ is contained in a commutative C*-subalgebra, $\mathcal{Z}$, of $\mathcal{L}(\mathcal{H})$ generated by projections. Then $\pi(\mathcal{Z})$ is generated by projections and contains $\operatorname{im} \tau$, so Theorem 5.8 applies.

5.11 Remark. In the case $X \subset \mathbb{C}$, Corollary 5.10 says that if $n \in \mathcal{A}$ can be written in the form $\pi(N)$ with N normal in $\mathcal{L}(\mathcal{H})$, then N can be chosen so that $\sigma(N) = \sigma(n)$. Thus Corollary 5.10 can be regarded as a generalization of that result, the self-adjoint case of which was indirectly used in the proof.

5.12 Remark. The restriction to metric spaces for extensions is reasonable in light of the necessity of working on separable Hilbert spaces. As an interesting example of the sort of thing that can happen if this restriction is dropped, consider the following extension of $\mathcal{K}$ by $C(\beta N \backslash N)$, where βN is the Stone-Čech compactification of the natural numbers N. Let $\{\phi_n\}$ be an orthonormal basis for $\mathcal{H}$, let $\mathcal{E}$ consist of all compact perturbations of operators $D = \text{diag}(\lambda_n)$ diagonal relative to $\{\phi_n\}$, and define $\phi : \mathcal{E} \to C(\beta N \backslash N)$ by taking $\text{diag}(\lambda_n)$ to the restriction to $\beta N \backslash N$ of the extension to βN of the continuous bounded function (λ_n) on N. The extension $(\mathcal{E}, \phi)$ appears to be trivial but is not. In fact, there is no trivial extension of $\mathcal{K}$ by $C(\beta N \backslash N)$ on a separable Hilbert space because $\beta N \backslash N$ contains a continuum of pairwise disjoint clopen sets. Moreover, this example shows that metrizability of X cannot be omitted from the hypothesis of Theorems 5.8 and 5.10. Corollary 5.4 is also false for non-separable algebras.

§6. The First Splitting Lemma

This section is devoted to a splitting lemma and its (eventual) consequences. In particular, a rudimentary Mayer-Vietoris sequence is obtained that will play an important part in showing that Ext is always a group. We are indebted to B. Mackichan for the suggestion that such a sequence might be relevant to this problem.

<u>6.1 Lemma</u>. (First splitting lemma). Let $p : X \to Y$ be surjective and let $\tau : C(X) \to A$ be an extension such that $p_*(\tau)$ is trivial, so that there exists a projection-valued measure E on Y such that

$$\tau(g \circ p) = \pi \int_Y g dE \quad , \text{ for all } \quad g \in C(Y) \ .$$

If C is a closed subset of Y and ∂C is disjoint from the set of points in Y with multiple preimage in X, then $\pi E(C)$ commutes with $\operatorname{im} \tau$.

<u>Proof</u>. Let $A = p^{-1}(\partial C)$, $U = p^{-1}(C^o)$, and $V = p^{-1}(Y \setminus C)$. We first note that it is enough to prove that $\pi E(C)$ commutes with $\tau(f)$ when f vanishes on A. In fact, since p is 1-1 on A, for any $f \in C(X)$ there is $g \in C(Y)$ such that $g \circ p$ agrees with f on A. Since $\pi E(C)$ commutes with $\tau(g \circ p)$, we may replace f with $f - g \circ p$.

Now since any f vanishing on A can be approximated uniformly by functions vanishing on a neighborhood of A, we may assume f vanishes on a neighborhood of A. Then f is the sum of a function f_1 supported in U and a function f_2 supported in V. If f_1 has support on a compact subset K_1 of U, let $g_1 \in C(Y)$ be such that $g_1 = 1$ on $p(K_1)$ and g_1 has support in C^o. Then, since $f_1 = f_1 \cdot (g_1 \circ p) = (g_1 \circ p) \cdot f_1$,

$$\pi E(C)\tau(f_1) = [\pi E(C)\tau(g_1 \circ p)]\tau(f_1) = \tau(g_1 \circ p)\tau(f_1) = \tau(f_1)$$

and $\quad \tau(f_1)\pi E(C) = \tau(f_1)[\tau(g_1 \circ p)\pi E(C)] = \tau(f_1)\tau(g_1 \circ p) = \tau(f_1)$.

Similarly, $\pi E(C)\tau(f_2) = \tau(f_2)\pi E(C) = 0$. Therefore $\pi E(C)$ commutes with $\tau(f)$.

We want to thank Roger Howe for a suggestion which resulted in the simplified proof given for this lemma.

<u>6.2 Corollary</u>. If $X = \tilde{B} \cup \tilde{C}$ and $\tilde{B} \cap \tilde{C} = \{x_0\}$ for some $x_0 \in X$, then $\beta : \mathrm{Ext}(\tilde{B}) \oplus \mathrm{Ext}(\tilde{C}) \to \mathrm{Ext}(X)$ is an isomorphism.

<u>Proof</u>. Let $p : X \to [-1,1]$ be such that $p < 0$ on $\tilde{B} \setminus \{x_0\}$, $p > 0$ on $\tilde{C} \setminus \{x_0\}$, and $p(x_0) = 0$. Let $Y = p(X)$ and $C = Y \cap [0,1]$. Now if $\tau : C(X) \to \mathcal{A}$ is an element of $\mathrm{Ext}(X)$, then $p_*(\tau)$ is trivial in $\mathrm{Ext}(Y)$. Thus if $e = \pi E(C)$ as in Lemma 6.1, e commutes with $\mathrm{im}\, \tau$. As in Remark 5.9, e splits τ into b' and c' where $b' \in \mathrm{Ext}(B')$, $c' \in \mathrm{Ext}(C')$, $B' \subset \tilde{B}$ and $C' \subset \tilde{C}$. Then if $i : B' \to \tilde{B}$ and $j : C' \to \tilde{C}$ are the inclusions and $\tilde{b} = i_*(b')$ and $\tilde{c} = i_*(c')$, $[\tau] = \beta(\tilde{b},\tilde{c})$. Therefore β is onto.

To show that β is 1-1 we note that there is a retraction $r : X \to \tilde{B}$, defined by setting $r(x) = x_0$ for $x \in \tilde{C}$, and a similar retraction $s : X \to \tilde{C}$. Clearly (r_*, s_*) is a left inverse for β.

<u>6.3 Remark</u>. In [29] Olsen has proved that if A and B are in $\mathcal{L}(\mathcal{H})$ and AB is compact, then there is a projection P such that AP and $(1-P)B$ are compact. This result could have been used instead of Lemma 6.1 in the proof of Corollary 6.2. Alternatively Olsen's result can be interpreted in terms of our theory. In fact, Olsen reduces to the case where A and B are positive by replacing A by A^*A and B by BB^*. Then $T = A + iB$ is essentially normal and has essential spectrum homeomorphic to a subset of $\mathbb{R}$ ($\mathbb{R}$ is homeomorphic to the union of the positive coordinate axes). Hence T is a compact perturbation of a normal operator N, and P can be taken to be one of the spectral projections of N. Of course, Olsen's result is also a special case of Corollary 6.2.

If X is a totally disconnected compact metric space, then $C(X)$ is generated by its projections (characteristic functions), and it follows from Theorem 5.8 that $\mathrm{Ext}(X) = 0$ (this also follows directly from the Weyl-von Neumann Theorem, since such spaces can be embedded in $\mathbb{R}$). The next lemma is an extension of this fact. The notation X/A, where A is a closed subset of X, denotes the quotient space obtained by collapsing A to a point.

6.4 Lemma. If A is a closed subset of X such that X/A is totally disconnected, then the map $i_* : \text{Ext}(A) \to \text{Ext}(X)$ induced by the inclusion map $i : A \to X$ is an isomorphism.

Proof. Write X as the disjoint union of A and clopen sets $X_1, X_2, \ldots$ with $\text{diam}\, X_n \to 0$. Fix $a_n \in \partial A$ such that

$$\text{dist}(a_n, X_n) = \text{dist}(\partial A, X_n) .$$

Then the map $r : X \to A$, defined by $r|A = \text{id}$ and $r(x) = a_n$ for all $x \in X_n$, is continuous and satisfies $ri = \text{id}_A$, so $r_* i_* = \text{id}_{\text{Ext}(A)}$. We will show that $i_* r_* = \text{id}_{\text{Ext}(X)}$. To this end, fix a *-monomorphism $\tau : C(X) \to A(H)$.

We claim that there exist mutually orthogonal projections $E_n \in \tau(\chi_n)$, where χ_n is the characteristic function of X_n, such that with $E_0 = I - \sum_{n=1}^{\infty} E_n$,

(1) $\tau(g \circ r) = \tau(g \circ r)\pi(E_0) + \pi(\sum_{n=1}^{\infty} g(a_n)E_n)$ for all $g \in C(A)$

(2) $\tau' : g \to \tau(g \circ r)\pi(E_0)$ is a *-isomorphism.

We show first how to complete the proof on this hypothesis, and then we verify the claim.

It is required to produce a *-monomorphism $\sigma : C(X) \to L(H)$ such that τ is equivalent to $\tau \circ r^* \circ i^* + \pi\sigma$; that is,

(3) $\tau(f) \cong \tau(f \circ r) \oplus \pi\sigma(f)$

simultaneously for all $f \in C(X)$ (actually a *-homomorphism will do by Theorem 5.3). The decomposition $f = (f - f \circ r) + (f \circ r)$ shows that $C(X)$ is the linear direct sum of the ideal $Z(A)$ of functions vanishing on A, and the subalgebra $r^*C(A)$. Since each X_n is totally disconnected, there exists a *-isomorphism $\sigma_1 : Z(A) \to L((I-E_0)H)$ such that

$$(4) \quad \pi(0 \oplus \sigma_1) = \tau | Z(A) \quad \text{and} \quad \sigma_1(\chi_n) = E_n \ , \quad n \geq 1 \ ,$$

where 0 is the zero map into $L(E_0 H)$. We also define $\sigma_2 : r^*C(A) \to L((I-E_0)H)$ by

$$\sigma_2(g \circ r) = \sum_{n=1}^{\infty} g(a_n)E_n \quad ,$$

and then let $\sigma(f) = \sigma_1(f - f \circ r) + \sigma_2(f \circ r)$. This map is clearly *-linear; in order that it be a homomorphism it is necessary and sufficient that

$$\sigma_1((g \circ r)f) = \sigma_2(g \circ r)\sigma_1(f)$$

for all $f \in Z(A)$, $g \in C(A)$. For $f \in Z(A)$ the expansion $f = \Sigma f\chi_n$ converges in norm, so by linearity and continuity it is enough to establish the above relation for f's satisfying $f\chi_n = f$. Then $(g \circ r)f = g(a_n)f$ so

$$\sigma_1((g \circ r)f) = g(a_n)\sigma_1(f) \ ;$$

on the other hand $\sigma_1(f) = \sigma_1(f)\sigma_1(\chi_n) = \sigma_1(f)E_n$, so

$$\begin{aligned} \sigma_2(g \circ r)\sigma_1(f) &= \sigma_2(g \circ r)E_n\sigma_1(f) \\ &= g(a_n)E_n\sigma_1(f) = g(a_n)\sigma_1(f) \ . \end{aligned}$$

To show that σ has the required property (3), we will show that

$$\tau(f) \cong 0 \oplus \pi\sigma_1(f)$$

$$\tau(g \circ r) \cong \tau(g \circ r) \oplus \pi\sigma_2(g \circ r)$$

simultaneously for all $f \in Z(A)$ and $g \in C(A)$. The claims (1) and (2) mean that $r_*(\tau) \cong \tau'$; if $V : H \to E_0 H$ implements this equivalence, then $V \oplus I : H \oplus (1-E_0)H \to E_0 H \oplus (1-E_0)H$ converts the desired relations into

$$\tau(f) \cong 0 \oplus \pi\sigma_1(f)$$

$$\tau(g \circ r) \cong \tau(g \circ r)\pi(E_0) \oplus \pi\sigma_2(g \circ r) ,$$

and these are actually equalities by (4) and (1).

To establish (1), it is sufficient to find projections $\{E_n\}$ such that (1) holds for a sequence $\{g_m\}$ dense in $C_{\mathbb{R}}(A)$. Choose self-adjoint $H_m \in \tau(g_m \circ r)$ and mutually orthogonal projections $F_n \in \tau(\chi_n)$. Then

$$\pi(H_m F_n - g_m(a_n)F_n) = \tau((g_m \circ r)\chi_n - g_m(a_n)\chi_n) = 0 ,$$

so that $H_m F_n - g_m(a_n)F_n$ is compact for all m,n. In particular $[H_m, F_n]$ and $F_n H_m F_n - g_m(a_n)F_n$ are compact, so if $F_n^{(k)}$ is a projection of codimension k in F_n such that $F_n^{(k)} \to 0$ strongly as $k \to \infty$, it follows that

$$||[H_m, F_n^{(k)}]|| \to 0 \quad \text{as} \quad k \to \infty$$

for all m,n. Hence for each n there exists a projection F_n' of finite codimension in F_n such that

$$||[H_m, F_n']|| \leq \frac{1}{n^2} \quad \text{for all} \quad m \leq n .$$

Consider the matrix of H_m relative to the decomposition $H = \sum_{n=0}^{\infty} \oplus F_n' H$, where $F_0' = I - \sum_{n=1}^{\infty} F_n'$. All entries above the diagonal are compact, since

$$F_k' H_m F_n' = F_k'[H_m, F_n']F_n' \quad \text{for all} \quad n > k \geq 0 .$$

With the exception of the finite set of entries with $n < m$, we have

$$||F_k' H_m F_n'|| \leq \frac{1}{n^2} \quad ,$$

and it follows that the operator formed of the entries of H_m above the diagonal and 0's elsewhere is compact (because these entries are compact and norm square-summable). Hence

$$H_m = \sum_{n=0}^{\infty} \oplus H_m^{(n)} + \text{compact} \quad ,$$

where $H_m^{(n)} = F_n' H_m | F_n' \mathcal{H}$. Moreover

$$H_m^{(n)} = g_m(a_n) I_n + \text{compact}, \; n \geq 1$$

where I_n is the identity on $F_n' \mathcal{H}$, and consequently $H_m^{(n)}$ has a reducing projection $F_n^{(m)}$ of finite codimension such that

$$H_m^{(n)} = g_m(a_n) F_n^{(m)} + K_n^{(m)} + H_m^{(n)} (I - F_n^{(m)})$$

with $K_n^{(m)}$ compact and $||K_n^{(m)}|| \to 0$ as $n \to \infty$. For each n let F_n'' be the projection on the intersection of the ranges of $F_n^{(1)}, \ldots, F_n^{(n)}$. Then F_n'' is of finite codimension in F_n and

$$H_m = H_m' \oplus \sum_{n=1}^{\infty} g_m(a_n) F_n'' + \text{compact}, \; m \geq 1$$

relative to $\mathcal{H} = (I - \Sigma F_n'')\mathcal{H} \oplus (\Sigma F_n'')\mathcal{H}$. Finally let E_n be of codimension 1 in F_n'' for all $n \geq 1$, and let $E_0 = I - \Sigma E_n$. Then $E_n \in \tau(\chi_n)$ for $n \geq 1$ and (1) is satisfied for all g_m , hence for all $g \in C(A)$.

To see that 2) is satisfied, observe first that $\tau(g \circ r)$ commutes with $\pi(E_0)$ for $g = g_m$ by construction, and hence for all g . Therefore τ' is a *-homomorphism. Moreover, the final dropping-down from F_n'' to E_n implies that the spectrum of

$\tau(g_m \circ r)\pi(E_0)$ contains the cluster points of $\{g_m(a_n)\}$, and hence that

$$\overline{\lim_{n\to\infty}} |g_m(a_n)| \leq ||\tau(g_m \circ r)\pi(E_0)|| .$$

It follows that this relation holds for all real g. Suppose $k \in \ker \tau'$; without loss of generality k is real-valued. Then $k(a_n) \to 0$, so $f = \Sigma k(a_n)\chi_n$ and $h = k \circ r - f$ are continuous. But $h = 0$ outside A so $h = 0$ on ∂A ; since $f = 0$ on ∂A it follows that k vanishes there and $k(a_n) = 0$ for all n. Then from (1) follows $k = 0$.

For the next lemma we return to the context of Lemma 6.1: $p : X \to Y$ is a surjection, B is a closed subset of Y containing all points with multiple preimage in X, $A = p^{-1}(B)$, and we have the diagram

$$\begin{array}{ccc} X & \overset{p}{\to} & Y \\ i\uparrow & & \uparrow j \\ A & \underset{p'}{\to} & B \end{array}$$

where $p' = p|A$ and i,j are the inclusion maps.

<u>6.5 Lemma</u>. $\ker p_* \subset i_*(\ker p'_*)$.

<u>Proof</u>. We are given $\tau : C(X) \to \mathcal{A}$ such that $p_*\tau$ is trivial, so that

$$\tau(g \circ p) = \pi \int_Y g\,dE \quad , \quad g \in C(Y)$$

for some projection-valued measure E on Y. Let $\{U_n\}$ be a basis for the topology of $X\setminus A$ such that $C_n = \overline{U}_n$ is disjoint from A for all n, and let $\mathcal{M}$ be the C*-subalgebra of $\mathcal{A}$ generated by $\operatorname{im} \tau$ and all $\pi E(p(C_n))$, so that $\mathcal{M}$ is commutative by Lemma 6.1. This situation gives rise to a compact metric space $\tilde{X}$, a surjection

$u : \tilde{X} \to X$, and a *-isomorphism $\tilde{\tau} : C(\tilde{X}) \to M$ such that $u_*(\tilde{\tau}) = \tau$. We claim that 1) each $a \in A$ has a unique preimage in $\tilde{X}$ (so that u is a homeomorphism on $u^{-1}(A)$) , and 2) $\tilde{X} \backslash u^{-1}(A)$ is totally disconnected. Note first that (regarding the points of $\tilde{X}$ as complex homomorphisms of M) if $u(x) \notin C_n$ then $x(\pi E(p(C_n))) = 0$. For $u(x) \notin C_n$ implies $p(u(x)) \notin p(C_n)$, so that there exists $g \in C(Y)$ with $g(p(u(x))) = 1$ and $g = 0$ on $p(C_n)$, and therefore

$$\begin{aligned} 0 &= x(\tau(g \circ p)\pi E(p(C_n))) \\ &= g(p(u(x)))x(\pi E(p(C_n))) \\ &= x(\pi E(p(C_n))) . \end{aligned}$$

In particular, if $u(x) \in A$ then x annihilates all $\pi E(p(C_n))$, and 1) is established. To simplify the notation we now identify $u^{-1}(A)$ with A .

In the same way $u(x) \in U_n$ implies that $x(\pi E(p(C_n))) = 1$. For $x_1, x_2 \notin A$ with $u(x_1) \neq u(x_2)$, it follows that x_1 and x_2 are distinguished by some $\pi E(p(C_n))$, and hence that x_1 and x_2 are separated by a clopen set. If $u(x_1) = u(x_2)$, then $x_1 = x_2$ on im τ, so that either they are again distinguished by some $\pi E(p(C_n))$, or else $x_1 = x_2$. This demonstrates 2).

Now let $k : A \to \tilde{X}$ be inclusion, let $r : \tilde{X} \to A$ be the retraction of Lemma 6.4, and let $\tau' = r_*(\tilde{\tau})$. Then from Lemma 6.4 we get

$$i_*(\tau') = i_* r_*(\tilde{\tau}) = u_* k_* r_*(\tilde{\tau}) = u_*(\tilde{\tau}) = \tau .$$

To complete the proof it must be shown that $p_*^!(\tau')$ is trivial. For this let N be the C*-subalgebra generated by im $p_*(\tau)$ and the $\pi E(p(C_n))$, and consider the commutative diagram

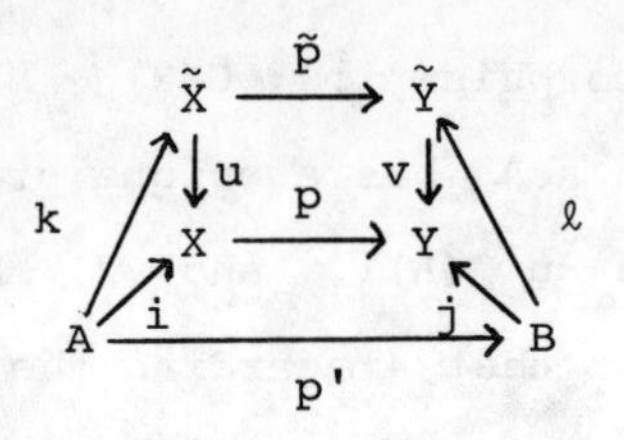

where the upper rectangle is dual to

$$\begin{array}{ccc} M & \supset & N \\ \cup & & \cup \\ \operatorname{im} \tau & \supset & \operatorname{im} p_*(\tau) \end{array}$$

and ℓ is inclusion (v is a homeomorphism on $v^{-1}(B)$ in the same way that u is on $u^{-1}(A)$). Now N is contained in the C*-algebra generated by the projections $\pi(\operatorname{im} E)$, so $\tilde{p}_* k_*(\tau')$ is trivial by Theorem 5.8. Hence $\ell_* p'_*(\tau')$ is trivial. But ℓ_* is an isomorphism by Lemma 6.4 (since $\tilde{Y} \setminus v^{-1}(B)$ is totally disconnected), and therefore $p'_*(\tau')$ is trivial.

We are now able to obtain several important results concerning exactness. In stating them, an assertion that a sequence

$$G \underset{p}{\rightarrow} H \underset{q}{\rightarrow} K$$

is exact will mean that $\operatorname{im} p = \ker q$. Since $\operatorname{Ext}(X)$ has not yet been shown to be a group, there is an a priori distinction between this and the usual notion of exactness (two elements of H that coincide in K differ by an element coming from G).

<u>6.6 Theorem</u>. If A is a closed subset of X, then

$$\operatorname{Ext}(A) \underset{i_*}{\rightarrow} \operatorname{Ext}(X) \underset{p_*}{\rightarrow} \operatorname{Ext}(X/A)$$

is exact, where $i : A \to X$ is the inclusion map and $p : X \to X/A$ is the quotient map.

Proof. The inclusion $\ker p_* \subset \operatorname{im} i_*$ follows from Lemma 6.5, and the other is obvious.

6.7 Theorem (Mayer-Vietoris sequence). Let B and C be closed subsets of X with $X = B \cup C$, and let i_1, i_2, j_1, j_2 be the inclusion maps of $B \cap C$ in B, $B \cap C$ in C, B in X, and C in X. Then

$$\operatorname{Ext}_1(B \cap C) \underset{\alpha}{\to} \operatorname{Ext}(B) \oplus \operatorname{Ext}(C) \underset{\beta}{\to} \operatorname{Ext}(X)$$

is exact, where Ext_1 denotes the group of invertible elements, and α and β are defined by

$$\alpha(a) = (i_{1*}(a), i_{2*}(-a)); \ a \in \operatorname{Ext}_1(B \cap C)$$

$$\beta(b,c) = j_{1*}(b) + j_{2*}(c); \ b \in \operatorname{Ext}(B), \ c \in \operatorname{Ext}(C) .$$

Proof. We have $\beta\alpha = 0$ since $j_1 \circ i_1 = j_2 \circ i_2$. Suppose $\beta(b,c) = 0$ for $b \in \operatorname{Ext}(B)$, $c \in \operatorname{Ext}(C)$. Using the isomorphism of Theorem 4.8 and Remark 5.9, it follows that $p_*(b \vee c) = 0$, where $p : B \vee C \to X$ is the obvious map. By Lemma 6.5, there exists $d \in \operatorname{Ext}(A \vee A)$ such that $i_*(d) = b \vee c$ and $p'_*(d) = 0$, where $A = B \cap C$, $i : A \vee A \to B \vee C$ is inclusion, and $p' : A \vee A \to A$ is the restriction of p. By Theorem 4.8 we get $d = a_1 \vee a_2$ for some a_1, $a_2 \in \operatorname{Ext}(A)$, and $a_1 + a_2 = p'_*(d) = 0$. Thus $a_1 \in \operatorname{Ext}_1(A)$ with inverse a_2. But

$$i_{1*}(a_1) \vee i_{2*}(a_2) = i_*(a_1 \vee a_2) = i_*(d) = b \vee c \quad ,$$

and therefore

$$\alpha(a_1) = (i_{1*}(a_1), i_{2*}(a_2)) = (b,c)$$

by Theorem 4.8 again.

6.8 Corollary. If Ext(X) is a group, then Ext(A) is a group for any closed subset A of X .

Proof. From Theorem 6.7 we have the exact sequence

$$\mathrm{Ext}_1(A) \xrightarrow{\alpha} \mathrm{Ext}(A) \oplus \mathrm{Ext}(X) \xrightarrow{\beta} \mathrm{Ext}(X) .$$

If $a \in \mathrm{Ext}(A)$ and $i : A \to X$ is inclusion, then $\beta(a,-i_*(a)) = 0$, and consequently there exists $a' \in \mathrm{Ext}_1(A)$ such that $\alpha(a') = (a,-i_*(a))$. That is, $(a',i_*(-a')) = (a,-i_*(a))$. In particular $a' = a$ and a is invertible.

§7. The Second Splitting Lemma

Let I^ω denote the Hilbert cube (the product of countably many copies of the unit interval). Since any compact metric space can be embedded in I^ω , it will follow from Theorem 6.6 that Ext is always a group, if it can be established that $\mathrm{Ext}(I^\omega)$ is a group. This we do in the next section, by iterating the "second splitting lemma" of the present section.

We begin with a slight variant of a classical lemma of von Neumann. Although this is undoubtedly known, we offer a proof for completeness.

7.1 Lemma. If H is a self-adjoint operator on $\mathcal{H}$, $\mathcal{F}$ is a finite-dimensional subspace of $\mathcal{H}$, and $\varepsilon > 0$, then there exist a finite-dimensional subspace $\mathcal{F}' \supset \mathcal{F}$ and a compact self-adjoint operator K such that $H+K$ is reduced by $\mathcal{F}'$ and $||K|| < \varepsilon$.

Proof. Let $\{\Delta_i\}$ be a decomposition of $\sigma(H)$ into a finite number of Borel sets of diameter less than ε , and let $\mathcal{F}' = \Sigma E(\Delta_i)\mathcal{F}$, where E is the spectral resolution of H . If P_i is the projection on $E(\Delta_i)\mathcal{F}$ and $P = \Sigma P_i$, it suffices to show that $||(I-P)HP|| < \varepsilon$ (for then $K = -(I-P)HP - PH(I-P)$ has the required properties). Fix $\lambda_i \in \Delta_i$; then

$$(I-P)HP = (I-P)\Sigma HP_i = (I-P)\Sigma(H-\lambda_i I)P_i$$

$$||(I-P)HP|| \le ||\Sigma(H-\lambda_i I)P_i||$$

$$\le \max ||(H-\lambda_i I)P_i||$$

$$\le \max ||(H-\lambda_i I)E(\Delta_i)|| < \varepsilon$$

since $\Sigma(H-\lambda_i I)P_i$ is essentially an orthogonal sum.

In the next lemma an operator matrix (A_{ij}) will be called <u>n-diagonal</u> if $A_{ij} = 0$ for $|i-j| > n$.

<u>7.2 Lemma</u>. For any self-adjoint operators $H_0, H_1, \ldots$ on $\mathcal{H}$, there exist compact self-adjoint operators $K_0, K_1, \ldots$ on $\mathcal{H}$ and a decomposition $\mathcal{H} = \sum_{k=0}^{\infty} \oplus \mathcal{H}_k$ into finite-dimensional subspaces relative to which the operator matrix for H_0+K_0 is diagonal and that for H_n+K_n is $(n+1)$-diagonal, $n \ge 1$.

<u>Proof</u>. We apply Lemma 7.1 recursively. Fix an orthonormal basis $\{\phi_{ij} | 0 \le j \le i < \infty\}$ for $\mathcal{H}$. At stage one choose a finite-dimensional subspace F_{00} containing ϕ_{00} , and a compact self-adjoint operator K_{00} , $||K_{00}|| < 1$, such that H_0+K_{00} is reduced by F_{00} . At stage two we choose finite-dimensional F_{10} containing $F_{00} + [\phi_{10}]$ and compact K_{10} , $||K_{10}|| < \frac{1}{4}$, such that $H_0+K_{00}+K_{10}$ is reduced by both F_{00} and F_{10} (apply Lemma 7.1 to $(H_0+K_{00})|F_{00}^{\perp}$); and we choose finite-dimensional F_{11} containing $F_{10} + [\phi_{11}]$ and compact K_{11} , $||K_{11}|| < \frac{1}{4}$, such that H_1+K_{11} is reduced by F_{11} . Iteration of this procedure, making n applications of Lemma 7.1 at the nth stage, produces finite-dimensional subspaces F_{ij} and compact self-adjoint operators K_{ij} $(0 \le j \le i < \infty)$ such that

i) $\phi_{ij} \in F_{ij}$

ii) $||K_{ij}|| < 1/(i+1)^2$

iii) $H_n + \sum_{m=n}^{\infty} K_{mn}$ is reduced by $\mathcal{F}_{mn}$, $m \geq n$

iv) $\mathcal{F}_{ij} \subset \mathcal{F}_{i,j+1}$ and $\mathcal{F}_{ii} \subset \mathcal{F}_{i+1,0}$

Put $K_n = \sum_{m=n}^{\infty} K_{mn}$ (compact since the series of norms is convergent) and $\mathcal{H}_k = \mathcal{F}_{k,0} \ominus \mathcal{F}_{k-1,0}$. Then $\mathcal{H} = \sum_{k=0}^{\infty} \oplus \mathcal{H}_k$ by i) and iv), and this decomposition reduces H_0+K_0 by iii). Moreover the inclusions

$$\mathcal{H}_k \subset \mathcal{F}_{k,0} \subset \mathcal{F}_{k+n,n} \subset \mathcal{F}_{k+n+1,0}$$

imply that

$$\mathcal{H}_k \subset \mathcal{F}_{k+n,n} \subset \mathcal{H}_0 \oplus \mathcal{H}_1 \oplus \ldots \oplus \mathcal{H}_{k+n+1}$$

and hence that (using iii) again)

$$(H_n+K_n)\mathcal{H}_k \subset \mathcal{H}_0 \oplus \ldots \oplus \mathcal{H}_{k+n+1}$$

for all k and n. Since H_n+K_n is self-adjoint it follows that it is $(n+1)$-diagonal.

7.3 Lemma. (Second Splitting Lemma). For any self-adjoint elements $h_0, h_1 \ldots$ of $A(\mathcal{H})$ such that h_0 commutes with all h_n there exists p in $A(\mathcal{H})$ such that $0 \leq p \leq 1$, p commutes with all h_n, $pf(h_0) = f(h_0)$ for all continuous f vanishing on $[\frac{1}{2}, +\infty)$, and $pf(h_0) = 0$ for all continuous f vanishing on $(-\infty, \frac{1}{2}]$.

Proof. By Lemma 7.2, there exist self-adjoint operators H_n with $\pi(H_n) = h_n$ and a decomposition $\mathcal{H} = \Sigma \oplus \mathcal{H}_k$ into finite-dimensional subspaces relative to which $H_0 = \Sigma \oplus H_{0k}$ is diagonal and H_n is $(n+1)$-diagonal, $n \geq 1$. We now construct a sequence of continuous functions $\phi_k : \mathbb{R} \to [0,1]$ such that

a) $\{\phi_k\}$ decreases to the characteristic function of $(-\infty, \frac{1}{2}]$ and vanishes on $[2,\infty)$

b) $||\phi_k - \phi_{k+1}||_\infty \to 0$

c) $\lim_{k\to\infty} ||[\phi_k(H_0), H_n]|\mathcal{H}_k|| = 0$, $n \geq 1$.

Then with $P = \Sigma\oplus\phi_k(H_{0k})$ it follows that $p = \pi(P)$ has the required properties. Obviously $0 \leq P \leq I$ and $[P,H_0]$ is compact. To see that $[P,H_n]$ is compact, we use the following simple observation: if (relative to $\mathcal{H} = \Sigma\oplus\mathcal{H}_k$) S is diagonal and T is n-diagonal, then $[S,T]$ is compact if and only if $||[S,T]|\mathcal{H}_k|| \to 0$ (for $[S,T]$ is the sum of its restrictions $[S,T]|\sum_{k=0}^{\infty}\oplus\mathcal{H}_{(n+1)k+q}$, $0 \leq q \leq n$, and each of these is essentially an orthogonal sum). Thus we want $||[P,H_n]|\mathcal{H}_k|| \to 0$. But

$$[P,H_n]|\mathcal{H}_k = [P-\phi_k(H_0),H_n]|\mathcal{H}_k + [\phi_k(H_0),H_n]|\mathcal{H}_k \quad ,$$

the second term tends to zero by c) , and the first term is dominated in norm by

$$2\sup\{||\phi_j-\phi_k||_\infty | k-n-1 \leq j \leq k+n+1\}\cdot||H_n||$$

(since H_n is (n+1)-diagonal), which tends to zero by b) . Moreover, if f vanishes on $[\frac{1}{2}, +\infty)$ then $\phi_k f = f$ for all k , so that $Pf(H_0) = f(H_0)$; and if f vanishes on $(-\infty, \frac{1}{2}]$ then $||\phi_k f||_\infty \to 0$ and therefore $Pf(H_0)$ is compact.

To construct the sequence $\{\phi_k\}$, let f_j be continuous, 1 on $(-\infty, \frac{1}{2}]$, 0 on $[\frac{1}{2}+\frac{1}{j}, +\infty)$, and linear in between. Then $\{f_j\}$ decreases to the characteristic function of $(-\infty, \frac{1}{2}]$ and $||f_j-f_{j+1}||_\infty \to 0$. Since $[H_0,H_n]$ is compact, it follows that $[f(H_0),H_n]$ is compact for all continuous f , and hence that $||[f(H_0),H_n]|\mathcal{H}_k|| \to 0$ by the observation above. Choose $N_1 < N_2 < \ldots$ such that

$$||[f_j(H_0),H_n]|\mathcal{H}_k|| \leq 1/j \quad \text{for } k \geq N_j \text{ and } n \leq j \ .$$

The sequence $\{\phi_k\}$ defined by $\phi_k = f_1$ for $k < N_1$ and $\phi_k = f_j$ for $N_j \leq k < N_{j+1}$ then has the required properties a) - c) .

7.4 Corollary. Let X be a closed subset of the unit square I^2 such that X contains $\{\frac{1}{2}\}\times I$, let $B = X\cap([0,\frac{1}{2}]\times I)$, and let $C = X\cap([\frac{1}{2},1]\times I)$. Then $\beta : \text{Ext}(B) \oplus \text{Ext}(C) \to \text{Ext}(X)$ (c.f. Remark 5.9) is surjective.

Proof. Let $[\tau] \in \text{Ext}(X)$ and let $n = \tau(\chi)$, so that n is normal with spectrum X and $\tau(f) = f(n)$. Let $n = h_0+ih_1$ with h_0,h_1 self-adjoint, and let p be as provided by Lemma 7.3 (with $h_n = h_1$ for $n \geq 1$). Then $\tau_0(f) = f(\frac{1}{2} + ih_1)$ is well-defined, since $\{\frac{1}{2}\}\times I \subset X$, and trivial, so $\tau' = \tau + \tau_0$ is equivalent to τ by Theorem 5.5 (or use Corollary 2.3). Now

$$p' = \begin{pmatrix} p & [p(1-p)]^{\frac{1}{2}} \\ [p(1-p)]^{\frac{1}{2}} & 1-p \end{pmatrix}$$

is a projection that commutes with $\text{im}\,\tau'$, so $\tau_1 = p'\tau'$ and $\tau_2 = (1-p')\tau'$ determine $b_1 \in \text{Ext}(B_1)$ and $c_1 \in \text{Ext}(C_1)$ for suitable closed subsets B_1 and C_1 of X (c.f. Remark 5.9). Moreover, $B_1 \subset B$, for if $f \in C[0,1]$ vanishes precisely on $[0,\frac{1}{2}]$ and $g \in C(X)$ is defined by $g(z) = f(x)$, then

$$\tau_1(g) = p'g(n \oplus (\tfrac{1}{2} + ih_1)) = p'(f(h_0) \oplus 0) = 0$$

by Lemma 7.3, and therefore $B_1 \subset \{g=0\} = B$. In the same way $C_1 \subset C$. Now let $i_1 : B_1 \to B$, $i : B \to X$, $j_1 : C_1 \to C$, and $j : C \to X$ be the inclusion maps. Then $i\circ i_1$ and $j\circ j_1$ are the inclusions of B_1 and C_1 in X , so $(i\circ i_1)_*(b_1) + (j\circ j_1)_*(c_1) = [\tau'] = [\tau]$ by construction. Hence if $b = i_{1*}(b_1)$ and $c = j_{1*}(c_1)$, then $b \in \text{Ext}\,B$, $c \in \text{Ext}\,C$, and $\beta(b,c) = i_*(b) + j_*(c) = [\tau]$.

<u>7.5 Corollary</u>. Let B and C be closed subsets of X with $B \cup C = X$, and let $A = B \cap C$. Let $\tilde{X} \subset X \times I$ be defined by

$$\tilde{X} = (B \times \{1\}) \cup (A \times I) \cup (C \times \{0\}) \quad ,$$

and let $f : \tilde{X} \to X$ be projection. Then f_* is surjective.

<u>Proof</u>. Let $\tau : C(X) \to \mathcal{A}$ be an extension, let $f_0 : X \to [0,1]$ be such that $f_0 < \frac{1}{2}$ on $B \setminus A$, $f_0 = \frac{1}{2}$ on A, and $f_0 > \frac{1}{2}$ on $C \setminus A$, let $\{f_1, f_2, \ldots\}$ be dense in $C_{\mathbb{R}}(X)$, let $h_n = \tau(f_n)$ for $n \geq 0$, and let p be given by Lemma 7.3. If $\mathcal{B}$ is the commutative C*-subalgebra generated by p and $\operatorname{im} \tau$, then there exist compact metric X', a surjection $s : X' \to X$, and a *-isomorphism $\tau' : C(X') \to \mathcal{B}$, such that $s_*(\tau') = \tau$. We show that there exists an injection $u : X' \to \tilde{X}$ such that $f \circ u = s$; then we will have $f_*(u_*(\tau')) = \tau$ as required. Consider $u : x' \to (x'|\operatorname{im} \tau, x'(p))$, so $u(x') \in X \times I$, elements of X and X' being regarded as complex homomorphisms of $\operatorname{im} \tau$ and $\mathcal{B}$. We need only show $u(x') \in \tilde{X}$ for all $x' \in X'$. This is obvious if $x'|\operatorname{im} \tau \in A$, so suppose that $x'|\operatorname{im} \tau \in B \setminus A$; that is, there exists $b \in B \setminus A$ such that $x'\tau(g) = g(b)$ for all $g \in C(X)$. Since $f_0(b) < \frac{1}{2}$, there exists $\phi : I \to I$ with $\phi(f_0(b)) = 1$ and $\phi = 0$ on $[\frac{1}{2}, 1]$, and then $p\phi(h_0) = \phi(h_0)$ by Lemma 7.3. This means that $p\tau(\phi \circ f_0) = \tau(\phi \circ f_0)$, and on applying x' we find $x'(p) = 1$. Therefore $u(x') \in B \times \{1\}$. Similarly, $x'|\operatorname{im} \tau \in C \setminus A$ implies $x'(p) = 0$ and hence $u(x') \in C \times \{0\}$.

<u>7.6 Remark</u>. We will not use Corollary 7.5 since Corollaries 6.2 and 7.4 are adequate for present purposes. However, it is Corollary 7.5 that reveals the geometric meaning of Lemma 7.3. The space $\tilde{X}$ is obtained from X by "stretching" A. The quotient space arising from $\tilde{X}$ by collapsing $B \times \{1\}$ and $C \times \{0\}$ to (different) points is SA, the suspension of A. Thus the obstruction to representing an element of $\operatorname{Ext}(X)$ as the sum of elements of $\operatorname{Ext}(B)$ and $\operatorname{Ext}(C)$ lies in $\operatorname{Ext}(SA)$.

§8. Infinite Sums and Projective Limits

In general the sum of an infinite number of extensions cannot be meaningfully defined, the difficulty being that it is not possible to regard $\Sigma\oplus A(H_n)$ as a subalgebra of $A(\Sigma\oplus H_n)$. In certain cases this problem can be avoided, and the resulting infinite sum will be useful.

<u>8.1 Definition.</u> If $\{X_n\}$ is a sequence of closed subsets of X such that $\mathrm{diam}(X_n) \to 0$, and $\tau_n : C(X_n) \to A(H_n)$ is an extension, $n \geq 1$, then $\Sigma\tau_n : C(\overline{\cup X_n}) \to A(\Sigma\oplus H_n)$ is defined as follows. Fix $x_n \varepsilon X_n$ for each n. Given $f \varepsilon C(\overline{\cup X_n})$, choose T_n such that $\pi(T_n) = \tau_n(f|X_n)$ and $||T_n - f(x_n)|| \to 0$. Then put $(\Sigma\tau_n)(f) = \pi(\Sigma\oplus T_n)$

This definition requires several remarks. First

$$||\tau_n(f|X_n - f(x_n))|| = ||(f|X_n) - f(x_n)||_\infty \to 0$$

since $\mathrm{diam}(X_n) \to 0$, and it follows that the required T_n's exist. Next, the definition is independent of the choice of x_n's and T_n's, for if $\{x_n'\}$, $\{T_n'\}$ is another such choice, then $T_n - T_n'$ is compact and

$$||T_n - T_n'|| \leq ||T_n - f(x_n)|| + ||T_n' - f(x_n')|| + |f(x_n) - f(x_n')| \to 0 ,$$

so $\Sigma\oplus T_n - \Sigma\oplus T_n'$ is compact. Finally, it is clear that $\Sigma\tau_n$ is a *-isomorphism.

The following property of this sum is required. If $p: X \to Y$ is continuous, $Y_n = p(X_n)$, and $p_n : X_n \to Y_n$ is the restriction of p, then $p'_*(\Sigma\tau_n) = \Sigma p_{n*}(\tau_n)$ (where $p' = p|\overline{\cup X_n}$).

<u>8.2 Theorem</u> If X is the union of closed subsets $\{X_n\}$ such that $\mathrm{diam}(X_n) \to 0$ and $X_m \cap X_n = \{x_0\}$ for some $x_0 \varepsilon X$ and all $m \neq n$, then the map $S: \prod_{n=1}^{\infty} \mathrm{Ext}(X_n) \to \mathrm{Ext}(X)$ defined by $S(\tau_1, \tau_2, \ldots) = \sum_{n=1}^{\infty} \tau_n$ is an isomorphism.

Proof. There is a retraction $r_n : X \to X_n$ defined by setting $r_n(x) = x_0$ for $x \notin X_n$. Since $r_{m*}(\Sigma\tau_n) = \tau_m$, it is clear that S is injective.

Now define closed sets Y_n by $Y_n = \bigcup_{m>n} X_m$, and let $\tau : C(X) \to A$ be an extension. Since $X = X_1 \cup X_2 \cup \ldots \cup X_n \cup Y_n$, by repeated application of Corollary 6.2 we obtain mutually orthogonal projections $p_1, p_2, \ldots \varepsilon A$ and extensions $\tau_n = p_n\tau$ and $\tau_n' = (1-p_1-p_2-\ldots-p_n)\tau$ so that $\tau = q_{n*}(\tau_1 \vee \tau_2 \vee \ldots \vee \tau_n \vee \tau_n')$. (Here $q_n : X_1 \vee X_2 \vee \ldots \vee X_n \vee Y_n \to X$ is the natural map and τ_n and τ_n' are obtained by splitting τ_{n-1}'.) There are orthogonal projections $P_n \varepsilon L(H)$ such that $\pi(P_n) = p_n$ and $\sum_{n=1}^{\infty} P_n = I$. If τ_n is considered to define an element of $\mathrm{Ext}(X_n)$ by mapping $C(X_n) \to A(P_n H)$, then we claim $\tau = \sum_{n=1}^{\infty} \tau_n$. In fact it is easy to see that $\tau(f) = (\sum_{n=1}^{\infty} \tau_n)(f)$ if f is constant on some Y_n, and such functions are dense in $C(X)$. Hence S is surjective and thus is an isomorphism.

8.3 Remark An easier related case occurs when X has disjoint clopen subsets X_n and a point x_0 such that $X \backslash \{x_0\} = \bigcup_{n=1}^{\infty} X_n$. In this case necessarily $\mathrm{diam}(X_n) \to 0$ and S is an isomorphism.

We turn now to a technical result on projective limits (c.f. [19]). Recall that if $\{X_n, f_n\}_{n \geq 1}$ is a sequence of compact spaces and continuous maps $f_n : X_{n+1} \to X_n$, then the projective limit (denoted proj. lim X_n) is a space X and "projection" maps $p_n : X \to X_n$ such that $f_n \circ p_{n+1} = p_n$ for all $n \geq 1$, and such that the following "universal property" holds: for any space Y and maps $q_n : Y \to X_n$ satisfying $f_n \circ q_{n+1} = q_n$, there exists a unique map $\phi : Y \to X$ such that $q_n = p_n \circ \phi$ for all $n \geq 1$. The projective limit is unique, and to prove existence, one may take

$$X = \{x \varepsilon \Pi X_n \mid f_n(x_{n+1}) = x_n \quad \text{for all} \quad n \geq 1\}$$

and $p_n(x) = x_n$. The projective limit of groups is defined analogously, with the maps required to be homomorphisms. We are interested in the relationship between $Ext(X)$ and proj. lim $Ext(X_n)$. There is an induced map $P:Ext(X) \to$ proj. lim $Ext(X_n)$, given by $P(a) = \{p_{n*}(a)\}$. Although this map is not injective in general [9], [19], we have:

8.4 Theorem. The induced map $P:Ext(X) \to$ proj. lim $Ext(X_n)$ is surjective.

Proof. Let $\{a_n\}$ define an element of proj. lim $Ext(X_n)$, so that $a_n \varepsilon Ext(X_n)$ and $a_n = f_{n*}(a_{n+1})$. Let $\tau_1 : C(X_1) \to A$ determine a_1 and $\tau_2' : C(X_2) \to A$ determine a_2, so that

$$\tau_2' \circ f_1^* + \tau_1^\circ \simeq \tau_1$$

for some trivial $\tau_1^\circ : C(X_1) \to A$. Thus there is a projection $e_1 \varepsilon A$ commuting with $im\, \tau_1$ such that $e_1\tau_1$ defines the trivial element of $Ext(X_1)$ and $(1-e_1)\tau_1 \simeq \tau_2' \circ f_1^*$. If $\alpha : A \to (1-e_1)A(1-e_1)$ is a *-isomorphism implementing this equivalence, then $\tau_2 = \alpha\tau_2'$ determines a_2 and $\tau_2 \circ f_1^* = (1-e_1)\tau_1$. Continuing in this manner, we construct an orthogonal sequence $\{e_n\}$ of projections in A and extensions $\tau_n : C(X_n) \to e_n' A e_n'$ (where $e_n' = 1-e_1-\ldots-e_{n-1}$) such that 1) τ_n determines a_n, 2) $e_n\tau_n$ defines the trivial element of $Ext(X_n)$, and 3) $e_{n+1}'\tau_n = \tau_{n+1} \circ f_n^*$.

Now let B be the C*-subalgebra of A generated by $\cup(im\tau_n)$ Note that B is commutative (by 3)) and contains all e_n. We calculate the maximal ideal space Y of B as follows. Let

$$Y_n = \{y \varepsilon Y | y(e_n) = 1 \text{ and } y(e_m) = 0 \text{ for } m \neq n\}$$

$$Y_\infty = \{y \varepsilon Y | y(e_n) = 0 \text{ for all } n \geq 1\} .$$

These sets are pairwise disjoint with union Y, the Y_n's are clopen, and Y_∞ is closed. For each $y \varepsilon Y_n$, $y|im(e_n\tau_n)$ is nontrivial and thus is given by evaluation at a point $\alpha_n(y)$ of X_n:

$$y(e_n\tau_n(g_n)) = g_n(\alpha_n(y)) \;,\; g_n \varepsilon C(X_n) \;.$$

It is easy to see that α_n is a homeomorphism. Next we show that Y_∞ is homeomorphic to X. If $y\varepsilon Y_\infty$ then $y|\mathrm{im}\tau_n$ is nontrivial, so there exists $\beta_n(y)\varepsilon X_n$ such that

$$y(\tau_n(g_n)) = g_n(\beta_n(y)) \;,\quad g_n\varepsilon C(X_n) \;.$$

Moreover $f_n \circ \beta_{n+1} = \beta_n$ for all $n \geq 1$, as can be seen from the computation

$$\begin{aligned} g_n(f_n(\beta_{n+1}(y))) &= y(\tau_{n+1}(g_n \circ f_n)) \\ &= y(e'_{n+1}\tau_n(g_n)) \\ &= y(\tau_n(g_n)) = g_n(\beta_n(y)) \end{aligned}$$

for all $g_n\varepsilon C(X_n)$ and $y\varepsilon Y_\infty$. Therefore $y \to \{\beta_n(y)\}$ defines a map $\phi: Y_\infty \to X$ which is clearly continuous and injective. To see that ϕ is surjective it must be shown that, given $x\varepsilon X$, there exists a simultaneous extension to $\mathcal{B}$ of the maps $\tau_n(g_n) \to g_n(x_n)$. Observe that each element of the *-algebra generated by $\bigcup \mathrm{im}\tau_n$ has a unique representation of the form

$$e_1\tau_1(g_1) + \ldots + e_n\tau_n(g_n) + \tau_{n+1}(g_{n+1})$$

for some $n \geq 1$ and $g_i\varepsilon C(x_i)$, and that the map taking this to $g_{n+1}(x_{n+1})$ determines a unique element of Y_∞ that extends all the above maps (since $\tau_n(g_n) = e_n\tau_n(g_n) + \tau_{n+1}(g_n \circ f_n)$ by 3) and $(g_n \circ f_n)(x_{n+1}) = g_n(x_n)$).

Next let $\tau: C(Y) \to \mathcal{B}$ be the inverse of the Gelfand transform. Let $g_n\varepsilon C(X_n)$ and define $\bar{g}_n\varepsilon C(Y)$ by $\bar{g}_n = g_n \circ \alpha_n$ on Y_n and $\bar{g}_n = 0$ elsewhere. Then for all $y\varepsilon Y_n$,

$$y\tau(\bar{g}_n) = \bar{g}_n(y) = g_n(\alpha_n(y)) = y(e_n\tau_n(g_n)) \;,$$

and since the extremes vanish for all $y \notin Y_n$, we find that

$\tau(\bar{g}_n) = e_n\tau_n(g_n)$, and hence that $\tau|C(Y_n)$ is trivial. It follows from this that $q_*(\tau)$ is trivial, where $q:Y \to Y/Y_\infty$ is the quotient map. In fact, if $\{E_n\}$ is an orthogonal sequence of projections in $L(H)$ with $\Sigma E_n = I$ and $\pi(E_n) = e_n$, and if $\sigma_n:C(Y_n) \to L(E_nH)$ trivializes $\tau|C(Y_n)$, then $\sigma = \Sigma\oplus\sigma_n$ trivializes $q_*(\tau)$. This is equivalent to the assertion that

$$\pi(\Sigma\oplus\sigma_n(f|Y_n)) = \tau(f)$$

for all $f\varepsilon C(Y)$ such that $f|Y_\infty$ is constant. If $f = c$ on Y_∞ , the latter assertion results from the fact that $\Sigma(f-c)\chi_n$ converges in norm to $f-c$, where χ_n is the characteristic function of Y_n .

Now by Theorem 6.6 there exists $a\varepsilon \mathrm{Ext}(Y_\infty)$ such that $i_*(a) = [\tau]$ where $i:Y_\infty \to Y$ is inclusion. The proof is completed by showing that $P(\phi_*(a)) = \{a_n\}$, or equivalently, that $\beta_{n*}(a) = a_n$ for all $n \geq 1$. Let $Y_n' = Y\setminus Y_1\cup\ldots\cup Y_{n-1}$, and let $\lambda_n:Y_n' \to X_n$ be defined by $y \to y|\mathrm{im}\tau_n$. If $\tau_n' = \tau|C(Y_n')$, it follows that $\lambda_{n*}(\tau_n') = \tau_n$. Moreover, if $j_n:Y_\infty \to Y_n'$ is inclusion, it follows as above that $j_{n*}(a) = [\tau_n']$. Since $\beta_n = \lambda_n \circ j_n$, we get

$$\beta_{n*}(a) = \lambda_{n*}[\tau_n'] = [\tau_n] = a_n$$

and the proof is complete.

<u>8.5 Remark</u>. The proof is much simpler when the maps f_n are all surjective (because the e_n can be taken to be zero). This is the situation in the application of the next section. However the proof of the surjectivity of γ in Theorem 10.5 requires the case in which all f_n are injective, so that $X_{n+1} \subset X_n$ and $X = \cap X_n$. The proof here is the same as for the general case, but is easier to follow.

§9. Proof that Ext is a Group

The main result of this section is the fact that Ext is always a group. Our proof is very indirect, even for planar sets, and a more direct approach would certainly be of interest.

9.1 Lemma. $\mathrm{Ext}(I^{\omega})$ is a group.

Proof. Let $a = [\tau] \varepsilon \mathrm{Ext}(I^{\omega})$, let $g_1, g_2, \ldots$ in $C(I^{\omega})$ be the coordinate maps, let $h_n = \tau(g_n)$, let

$$B = [0,\tfrac{1}{2}] \times \prod_{n=2}^{\infty} I \quad , \quad C = [\tfrac{1}{2},1] \times \prod_{n=2}^{\infty} I \quad ,$$

and apply Lemma 7.3 (with h_1 playing the rôle of h_0) just as in the proof of Corollary 7.4 to conclude that there exist $a' \varepsilon \mathrm{Ext}(I^{\omega})$ and $b \varepsilon \mathrm{Ext}(B \vee C)$ with $f_*(b) = a + a'$, where $f: B \vee C \to I^{\omega}$ is the obvious map. Since B and C are homeomorphic to I^{ω}, we can repeat this argument for each of the "pieces" of b (relative to the isomorphism between $\mathrm{Ext}(B \vee C)$ and $\mathrm{Ext}(B) \oplus \mathrm{Ext}(C)$). Iteration then produces the following situation:

i) a sequence $f_n : X_{n+1} \to X_n$ $(n \geq 1)$ of spaces and continuous maps, where $X_1 = I^{\omega}$, X_n has 2^{n-1} components each homeomorphic to I^{ω}, and the diameter of the components tends to zero.

ii) $a_n, a'_n \varepsilon \mathrm{Ext} X_n$ such that $f_{n*}(a_{n+1}) = a_n + a'_n$, where $a_1 = a$ is given.

Since the diameter of the components tends to zero, we conclude that the projective limit X of $\{X_n, f_n\}$ is totally disconnected, and that there exist well-defined elements

$$b_n = a_n + a'_n + f_{n*}(a'_{n+1}) + f_{n*}f_{n+1*}(a'_{n+2}) + \ldots$$

of $\mathrm{Ext}\, X_n$ by Definition 8.1 (because each a'_k can be decomposed into 2^{k-1} pieces corresponding to the components of X_k). Moreover

$$f_{n*}(b_{n+1}) = f_{n*}(a_{n+1})+f_{n*}(a'_{n+1})+f_{n*}f_{n+1*}(a'_{n+2})+\ldots$$

$$= a_n+a'_n+f_{n*}(a'_{n+1})+\ldots$$

$$= b_n$$

by the remark at the end of Definition 8.1. Therefore $\{b_n\}$ is an element of proj. lim $\mathrm{Ext}(X_n)$, so by Theorem 8.4 there exists $c\varepsilon\mathrm{Ext}(X)$ such that $p_{n*}(c) = b_n$ for all $n\geq 1$. But $\mathrm{Ext}(X) = 0$ because X is totally disconnected, so $b_1 = 0$ and

$$a'_1+f_{1*}(a'_2)+f_{1*}f_{2*}(a'_2)+ \ldots$$

is the required inverse for a .

9.2 Theorem. For any compact metrizable space X , $\mathrm{Ext}(X)$ is a group.

Proof. Any such X can be regarded as a closed subset of I^ω , so the result follows from Corollary 6.8 and Lemma 9.1.

9.3 Theorem. If $\{X_n\}$ is a sequence of closed subsets of $[0,1]$, then $\mathrm{Ext}\,(\prod_{n=1}^{\infty} X_n) = 0$.

Proof. Let $X = \Pi X_n$. We repeat the splitting argument of Lemma 9.1, but with the difference that since $\mathrm{Ext}(X)$ is now known to be a group, it is no longer necessary to add anything to obtain a splitting. Precisely, we require the following: if

$$B = (X_1\cap[0,\tfrac{1}{2}])\times \prod_{n=2}^{\infty} X_n \quad , \quad C = (X_1\cap[\tfrac{1}{2},1])\times \prod_{n=2}^{\infty} X_n$$

and $p:B\vee C \to X$ is the obvious map, then p_* is surjective. If $\frac{1}{2}\notin X_1$, then $B\vee C$ is homeomorphic to X and there is nothing to prove. If $\frac{1}{2}\varepsilon X_1$, define $r:X \to C$ by $r(x_1,x_2,\ldots) = (y_1,x_2,\ldots)$, where y_1 is x_1 or $\frac{1}{2}$ according as $x_1 \geq \frac{1}{2}$ or $x_1 \leq \frac{1}{2}$. Also let $i:B \to X$, $j:C \to X$ be the inclusion maps and $q:X \to X/B$ the quotient map, so that

$$\operatorname{Ext}(B) \xrightarrow{i_*} \operatorname{Ext}(X) \xrightarrow{q_*} \operatorname{Ext}(X/B)$$

is exact by Theorem 6.6. Then $q \circ j \circ r = q$, so that for any $a \in \operatorname{Ext}(X)$

$$q_*(a - j_* r_*(a)) = 0$$

and thus $a - j_* r_*(a) = i_*(b)$ for some $b \in \operatorname{Ext}(B)$. But then

$$p_*(b \vee r_*(a)) = i_*(b) + j_* r_*(a) = a$$

and the proof is complete.

<u>9.4 Corollary</u>. $\operatorname{Ext}(I^n) = 0$ for $0 \leq n \leq \omega$.

In a description [16] of certain elementary results from this paper, we raised the question of the triviality of a certain extension of $\mathcal{K}$ by $C(\mathbb{D})$, where $\mathbb{D}$ is the closed unit disk. This extension has since been shown to be trivial by Deddens and Stampfli [14] after its generator had been shown to be quasitriangular by Pearcy and Salinas [31]. Both results follow from the Corollary.

§10. Bijectivity of γ for Plane Sets

In this section we determine $\operatorname{Ext}(X)$ for $X \subset \mathbb{C}$ by showing that in this case the map

$$\gamma_X : \operatorname{Ext}(X) \to \operatorname{Hom}(\pi^1(X), \mathbb{Z})$$

of §5 is bijective. We begin with injectivity; for this we again use the iterated splitting argument of Lemma 9.1 to show that $a = 0$ whenever $\gamma(a) = 0$. For this argument to be applicable we must show that 1) the hypothesis allows a to be split along any straight line L as in Corollary 7.4, but without assuming that $X \cap L$ is an interval, and that 2) the pieces inherit the hypothesis.

<u>10.1 Lemma</u>. If A is a closed subset of $[0,1]$ and $X = [0,1]/A$, then γ_X is injective.

<u>Proof</u>. By considering the components of the complement of A ,

we see that X is the union of a sequence of closed subsets X_n, each homeomorphic to a circle or an interval and with $\operatorname{diam} X_n \to 0$, and that there is $x_0 \varepsilon X$ such that $X_m \cap X_n = \{x_0\}$ for all $m \neq n$. Moreover X can be regarded as a subset of $\mathbb{C}$.

The lemma is true when X is an interval, since then $\operatorname{Ext}(X) = 0$, or a circle, by the remarks following Theorem 3.1. We reduce the general case to this situation as follows. Let $\tau : C(X) \to A(H)$ be an extension in $\ker \gamma$, and let $Y_n = \bigcup_{i>n} X_i$ (a closed subset). As in the proof of Theorem 8.2 we obtain mutually orthogonal projections P_n and extensions $\tau_n : C(X_n) \to A(P_n H)$, $\tau_n' : C(Y_n) \to A((I - \sum_{i=1}^{n} P_i)H)$ such that $\tau = \tau_1 \vee \ldots \vee \tau_n \vee \tau_n'$, for all $n \geq 1$. If $f \varepsilon C(X)$ is constant on Y_n, we have

$$\tau(f) = \tau_1(f|X_1) \oplus \ldots \oplus \tau_n(f|X_n) \oplus f(x_0) ,$$

and it follows that τ_n is in $\ker \gamma$ for all n, and hence that each τ_n is trivial. Let $\sigma_n : C(X_n) \to L(P_n H)$ be a trivializing map for τ_n, and for functions f that are constant on some Y_n, define

$$\sigma(f) = \sigma_1(f|X_1) \oplus \ldots \oplus \sigma(f|X_n) \oplus f(x_0) .$$

Such functions are dense in $C(X)$, and it follows that the continuous extension of σ trivializes τ.

We require two mapping properties of $\ker \gamma$. The next lemma gives one of these, and the other is trivial: $f_*(\ker \gamma_X) \subset \ker \gamma_Y$ for any $f : X \to Y$.

<u>10.2 Lemma</u>. If $X \subset \mathbb{C}$, B and C are the intersections of X with the closed half-planes determined by a straight line L, and $\beta : \operatorname{Ext}(B) \oplus \operatorname{Ext}(C) \to \operatorname{Ext}(X)$ is the usual map, then

$$\ker(\gamma_X \beta) \subset \ker \gamma_B \oplus \ker \gamma_C .$$

<u>Proof</u>. We are given $*$-isomorphisms τ_1 and τ_2, determining elements of $\operatorname{Ext}(B)$ and $\operatorname{Ext}(C)$, such that

$$\operatorname{ind}(\tau_1(f|B)\oplus\tau_2(f|C)) = 0$$

for all $f:X \to \mathbb{C}^*(=\mathbb{C}\setminus\{0\})$. It must be shown that $\operatorname{ind}\tau_1(g) = 0$ for all $g:B \to \mathbb{C}^*$. Extend g to $g':B\cup L \to \mathbb{C}^*$ by linearity on the components of $L\setminus B\cup L$, steering around $\{0\}$ when necessary; and let $f = g'\circ p$, where $p:C \to L$ is projection and $p|B = \mathrm{id}$. Then $f|C$ is null-homotopic so that $\operatorname{ind}\tau_2(f|C) = 0$. Hence

$$0 = \operatorname{ind}\tau_1(f|B) = \operatorname{ind}\tau_1(g)$$

as required.

<u>10.3 Lemma</u>. If X is a compact subset of $\mathbb{C}$, then γ_X is injective.

<u>Proof</u>. Let $a\in\operatorname{Ext}(X)$ with $\gamma(a) = 0$, let L,B,C be as in Lemma 10.2, and let J be any compact interval on L that contains $X\cap L$. We need the following spaces and inclusion maps:

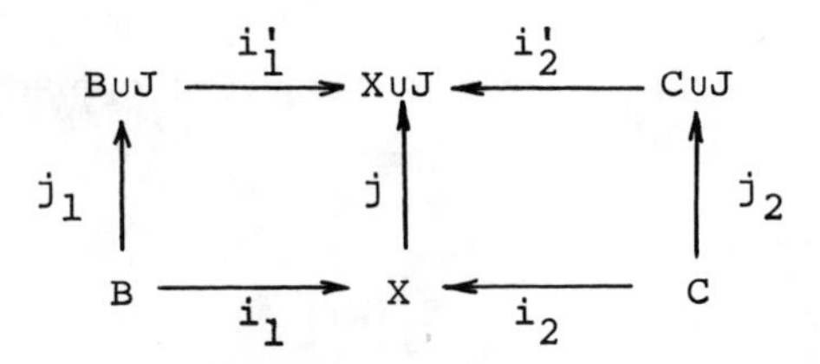

By Corollary 7.4 the map

$$\operatorname{Ext}(B\cup J)\oplus\operatorname{Ext}(C\cup J) \to \operatorname{Ext}(X\cup J)$$

is surjective, so there exist $b'\in\operatorname{Ext}(B\cup J)$ and $c'\in\operatorname{Ext}(C\cup J)$ with $j_*(a) = i'_{1*}(b')+i'_{2*}(c')$. Then $j_*(a)\in\ker\gamma$, so $b',c'\in\ker\gamma$ by Lemma 10.2, and therefore the images of b',c' in $\operatorname{Ext}(B\cup J/B)$ and $\operatorname{Ext}(C\cup J/C)$ are in $\ker\gamma$. These quotient spaces are homeomorphic to $J/B\cap C$, and hence by Lemma 10.1 the images of b',c' are trivial. But

$$\operatorname{Ext}(B) \to \operatorname{Ext}(B\cup J) \to \operatorname{Ext}(B\cup J/B)$$

is exact by Theorem 6.6, so there exist $b\in\operatorname{Ext}(B)$, $c\in\operatorname{Ext}(C)$ such that $b' = j_{1*}(b)$ and $c' = j_{2*}(c)$. The proof is completed by showing $\beta(b,c) = a$ (since then $b,c \in \ker \gamma$ by Lemma 10.2). But $j_*\beta(b,c) = j_*(a)$, so it suffices to show that j_* is injective. This follows from the Mayer-Vietoris sequence

$$\mathrm{Ext}(X \cap J) \to \mathrm{Ext}(X) \oplus \mathrm{Ext}(J) \to \mathrm{Ext}(X \cup J)$$

of Theorem 6.7, since $\mathrm{Ext}(X \cap J) = 0$.

<u>10.4 Lemma</u>. If X is a finite simplicial complex in $\mathbb{C}$, then γ_X is surjective.

<u>Proof</u>. If $f_1, \ldots, f_n : X \to S^1$ is a system of free generators for $\pi^1(X)$, it will suffice to produce extensions $\tau_1, \ldots, \tau_n$ such that $\mathrm{ind}\, \tau_i(f_i) = -1$ and $\mathrm{ind}\, \tau_i(f_j) = 0$ for $i \neq j$. To this end, choose $g_1, \ldots, g_n : S^1 \to X$ such that $f_i \circ g_j$ is homotopic to the identity map for $i = j$ and to the constant map for $i \neq j$, choose a normal operator N with essential spectrum X, and define $\tau_i(f) = f(\pi(T_{g_i} \oplus N))$, where T_{g_i} is the Toeplitz operator on H^2 associated with g_i. Then

$$\begin{aligned} \mathrm{ind}\, \tau_i(f_j) &= \mathrm{ind}(T_{f_j \circ g_i} \oplus f_j(N)) \\ &= \mathrm{ind}\, T_{f_j \circ g_i} \end{aligned}$$

is as required.

<u>10.5 Theorem</u>. If X is a compact subset of the plane, then γ_X is bijective.

<u>Proof</u>. Let C_n be a finite cover of X by closed discs of radius $1/n$, each meeting X, let Y_n be the union of the members of C_n, and let $X_n = Y_1 \cap \ldots \cap Y_n$. Then $X_1 \supset X_2 \supset \ldots$, $X = \cap X_n$, and each X_n is (homeomorphic to the underlying set of) a finite simplicial complex. Let $i_n : X_{n+1} \to X_n$ and $j_n : X \to X_n$ be the inclusion maps, and let a homomorphism $h : \pi^1(X) \to \mathbb{Z}$ be given. Then $hj_n^* : \pi^1(X_n) \to \mathbb{Z}$ is a homomorphism, so by Lemma 10.4 there exists $a_n \in \mathrm{Ext}\, X_n$ such that

$\gamma_n(a_n) = hj_n^*$ (where $\gamma_n = \gamma_{X_n}$) . We claim that $\{a_n\} \varepsilon$ proj.lim $\mathrm{Ext}(X_n)$. This means that $i_{n*}(a_{n+1}) = a_n$ for all n ; but since γ_n is injective (by Lemma 10.3) it suffices to have $\gamma_n i_{n*}(a_{n+1}) = \gamma_n(a_n)$, and this results from the commutativity of the diagram

$$\begin{array}{ccc} \mathrm{Ext}(X_{n+1}) & \xrightarrow{\gamma_{n+1}} & \mathrm{Hom}(\pi^1(X_{n+1}),\mathbb{Z}) \\ \downarrow i_{n*} & & \downarrow i_{n*} \\ \mathrm{Ext}(X_n) & \xrightarrow[\gamma_n]{} & \mathrm{Hom}(\pi^1(X_n),\mathbb{Z}) \end{array}$$

(that is, γ is a natural transformation of functors). Then by Theorem 8.4 there exists $a\varepsilon\mathrm{Ext}(X)$ such that $j_{n*}(a) = a_n$ for all n . The proof is completed by showing $\gamma_X(a) = h$. For all n we have

$$j_{n*}(\gamma_X(a)) = \gamma_n(j_{n*}(a)) = \gamma_n(a_n) = hj_n^* .$$

But for any $f:X \to \mathbb{C}^*$, there exists an integer n such that f has a continuous extension $f_n:X_n \to \mathbb{C}^*$ (that is, $\pi^1(X)$ is the direct limit of the $\pi^1(X_n)$) , and therefore

$$\begin{aligned} \gamma_X(a)[f] &= \gamma_X(a)j_n^*[f_n] \\ &= j_{n*}(\gamma_X(a))[f_n] \\ &= hj_n^*[f_n] = h[f] \end{aligned}$$

as required.

An alternate proof of the surjectivity of γ_X for $X \subset \mathbb{C}$ can be given using recent results of Berger and Shaw [6]. The generator for each extension can be taken to be the direct sum of multiplication operators and their adjoints defined on various Hilbert spaces of analytic functions. We will have more to say about this in the following section.

The surjectivity of γ_X for homologically locally connected X has been obtained independently by Ekman [20]. The starting point for his investigation was Problem 6 of [13].

§11. Applications and Open Problems

In this section we describe some applications of our results, mostly to operator theory. The first amounts to a restatement of the main theorem.

11.1 Theorem. If T_1 and T_2 are essentially normal operators on $\mathcal{H}$, then a necessary and sufficient condition that T_1 be unitarily equivalent to some compact perturbation of T_2 is that T_1 and T_2 have the same essential spectrum Λ and $\mathrm{ind}(T_1-\lambda I) = \mathrm{ind}(T_2-\lambda I)$ for all $\lambda \notin \Lambda$.

Proof. The necessity was mentioned in the Introduction. For sufficiency let τ_1 and τ_2 be the extensions of $\mathcal{K}$ by $C(\Lambda)$ determined by T_1 and T_2. By Theorem 10.5 it is enough to show that $\gamma_\Lambda[\tau_1] = \gamma_\Lambda[\tau_2]$. Now $\pi^1(\Lambda)$ is the free abelian group with a generator for each bounded component of $\mathbb{C}\setminus\Lambda$. Moreover, if U is such a component and $\lambda \in U$, then the homotopy class containing the invertible function θ_λ, defined by $\theta_\lambda(z) = z-\lambda$ for $z \in \Lambda$, is the generator corresponding to U. Since

$$\gamma_\Lambda[\tau_1][\theta_\lambda] = \mathrm{ind}(T_1-\lambda I) = \mathrm{ind}(T_2-\lambda I) = \gamma_\Lambda[\tau_2][\theta_\lambda]$$

it follows that $\gamma_\Lambda[\tau_1] = \gamma_\Lambda[\tau_2]$, which completes the proof.

A special case of this is of particular interest.

11.2 Corollary. If T is an essentially normal operator on $\mathcal{H}$ such that $\mathrm{ind}(T-\lambda I) = 0$ for all λ not in the essential spectrum of T, then T is in

$$\mathcal{N}+\mathcal{K} = \{N+K \mid N \text{ normal, } K \text{ compact}\}$$

11.3 Corollary. If T is an essentially normal operator on $\mathcal{H}$ whose essential spectrum does not separate the plane, then T is in $\mathcal{N}+\mathcal{K}$.

Since the converse of Corollary 11.2 is (trivially) valid and index is continuous, we obtain

11.4 Corollary. The family $\mathcal{N}+\mathcal{K}$ is norm-closed.

A direct proof of this would be highly desirable. From the theory for higher dimensional spaces [9] it follows that the family $\{(N_1+K_1, N_2+K_2) \mid N_1, N_2$ normal, $[N_1,N_2] = 0$, K_1, K_2 compact$\}$ is not norm-closed. Thus a direct proof would have to fail for two variables.

There are numerous unanswered questions related to this result. First, is the map $\pi: \mathcal{N} \to \pi(\mathcal{N})$ (relatively) open? In more concrete terms, given $N \in \mathcal{N}$ and $\varepsilon > 0$, does there exist $\delta > 0$ such that for every $N' \in \mathcal{N}$ with $||\pi(N) - \pi(N')|| < \delta$ there is $K \in \mathcal{K}$ such that $N'+K \in \mathcal{N}$ and $||(N'+K) - N|| < \varepsilon$? One approach to this problem, as well as to a direct proof for the corollary, is the problem of approximation by normal operators on finite dimensional spaces. Equally important is the problem of estimating the distance to the normal operators of a compression of a normal operator to a subspace of finite codimension. A related question is the determination of the distance from T to the normal operators, for $T \in \mathcal{N}+\mathcal{K}$. If this distance were

$$\inf\{||K|| \mid T+K \in \mathcal{N}\},$$

it would follow that $\pi: \mathcal{N} \to \pi(\mathcal{N})$ is relatively open.

One complicating aspect of these questions is the difficulty of deciding when a compact perturbation of a normal operator is normal. The "converse" is also difficult, but the following is one solution: the difference $N-N'$ of normal operators is compact if and only if $\pi(P'(C)) \leq \pi(P(U))$ for all compact C and open U with $C \subset U$, where P and P' are the spectral measures of N and N'. To estimate $||N-N'||$ is more difficult, and to determine P' in terms of P is perhaps hopeless.

A related but more specific question concerns a pair of unilateral shifts, and asks for conditions that force their difference to be compact. Equivalently, denoting the unilateral shift by U_+, for which unitary operators W is $W^*U_+W - U_+$ compact? This problem is intimately related to the determination of the group of automorphisms on the C*-algebra generated by the shift.

If (E,ϕ) is an extension of K by $C(X)$, then there is a homomorphism $\xi : \mathrm{Aut}(E) \to \mathrm{Aut}(C(X)) = \mathrm{Homeo}(X)$ since an automorphism must preserve K. If α in $\mathrm{Aut}(E)$ induces the identity $\xi(\alpha)$ in $\mathrm{Homeo}(X)$, then α is defined by a unitary operator W on H such that the commutator $[W,A]$ is compact for A in E. To determine the range of ξ we observe that $\xi(\mathrm{Aut}(E))$ must preserve $\gamma(E,\phi)$ in $\mathrm{Hom}(\pi^1(X),\mathbb{Z})$. In fact, if γ is an isomorphism, then this is both necessary and sufficient. In particular, for $X \subset \mathbb{C}$ we obtain the exact sequence

$$0 \to S^1 \to \{W \in U(H) \mid [W,A] \in K \text{ for all } A \in E\}$$

$$\to \mathrm{Aut}(E) \to \{\delta \in \mathrm{Homeo}(X) \mid \delta^*\gamma(E,\phi) = \gamma(E,\phi)\} \to 0 ,$$

where δ^* is the automorphism defined on $\mathrm{Hom}(\pi^1(X),\mathbb{Z})$ by the automorphism δ_* defined on $\pi^1(X)$ by δ.

Applying this to the C*-algebra T generated by the unilateral shift U_+ on H, we obtain the exact sequence

$$0 \to S^1 \to \{W \in \mathcal{U}(H) \mid [W,U_+] \in \mathcal{H}\} \to \mathrm{Aut}(\mathcal{T}) \to \mathrm{Homeo}_+(S^1) \to 0 ,$$

where $\mathrm{Homeo}_+(S^1)$ denotes the orientation preserving homeomorphisms on S^1.

This approach to the study of the group of automorphisms on the C*-algebra generated by the shift was initiated by Berger and Coburn.

A related application of these ideas was suggested by an observation of Helton.

<u>11.5 Theorem</u>. If (E_1,ϕ_1) and (E_2,ϕ_2) are extensions of K by $C(X_1)$ and $C(X_2)$, then E_2 is *-isomorphic to a subalgebra of E_1 if and only if there exists a surjection $f:X_1 \to X_2$ and an integer $n>0$ such that

$$f_*((E_1,\phi_1)) = n(E_2,\phi_2) .$$

<u>Proof</u>. If $\theta:E_2 \to E_1$ is a *-monomorphism, then $\theta|K$ must have finite multiplicity n. The map θ induces an injection $\hat{\theta}:C(X_2) \to C(X_1)$, and hence $\hat{\theta}$ is dual to a surjection $f:X_1 \to X_2$. It is easily checked that $f_*((E_1,\phi_1)) = n(E_2,\phi_2)$, and that the converse holds. (We note that if one demands $\theta(E_2) \supset \ker \phi_1$, then one has the same result except that $n = 1$.)

We conclude this group of problems with a final question. Let $R \in L(H)$ be essentially normal, let $r(U) = URU^*$ for $U \in \mathcal{U}(H)$ (the unitary group on H), and let $\mathcal{R} = r(\mathcal{U}(H))$. Is the map $\pi:\mathcal{R} \to \pi(\mathcal{R})$ relatively open? Does it have the homotopy lifting property? Note that $\pi(\mathcal{R})$ is closed and that $\pi \circ r$ induces a one-to-one correspondence between $\pi(\mathcal{R})$ and the left coset space $\mathcal{U}(H)/I_R$, where

$$I_R = \{U \in \mathcal{U}(H) \mid [\pi(U),\pi(R)] = 0\} ,$$

a closed subgroup of $\mathcal{U}(H)$. However, this correspondence is not topological ($\pi \circ r$ is not open) unless the essential spectrum $\Lambda = \sigma(\pi(R))$ is finite. Moreover $\pi \circ r$ fails to have the homotopy lifting property, at least when Λ contains an arc. If R is normal with $\sigma(R) = \sigma(\pi(R))$, the set $\mathcal{R}$ can be enlarged (without changing $\pi(\mathcal{R})$) to the set of all normals with spectrum and essential spectrum Λ, and the same questions can be asked. If R is the shift then π is open on $\mathcal{R}$, and if $\mathcal{R}$ is enlarged to the set of isometries of index -1, then π is open and has the homotopy lifting property. A related question is discussed in [8].

Theorem 11.1 enables us to present models for essentially normal operators. This was done for operators with essential spectrum equal to a simple closed curve in Corollary 3.2. In the general case we use the following results of Berger and Shaw [6]. Let U be a bounded open subset of $\mathbb{C}$, let $A^2(U)$ consist of those functions in $L^2(U, dxdy)$ that are holomorphic on U, and let T_U be multiplication by z on $A^2(U)$. Then T_U is essentially normal, $\sigma(T_U) = \overline{U}$, $\sigma(\pi(T_U)) \subset \partial U$, $\operatorname{ind}(T_U - \lambda I) = -1$ for $\lambda \varepsilon U$, and

$$\operatorname{tr}[T_U^*, T_U] = \frac{1}{\pi} \operatorname{Area}(U) .$$

Fix a compact subset X of $\mathbb{C}$, let $U_1, U_2, \ldots$ be the bounded components of $\mathbb{C} \backslash X$, and let $n_1, n_2, \ldots$ be an assignment of integers to the components. If N is a normal operator with essential spectrum X, and $U^* = \{\bar{z} | z \varepsilon U\}$, then the operator

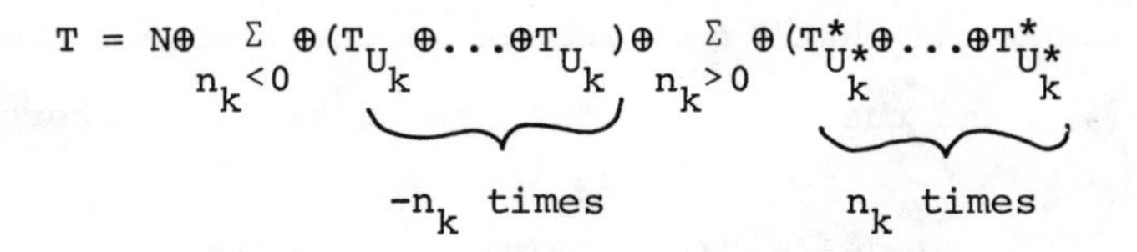

$$T = N \oplus \sum_{n_k < 0} \oplus (\underbrace{T_{U_k} \oplus \ldots \oplus T_{U_k}}_{-n_k \text{ times}}) \oplus \sum_{n_k > 0} \oplus (\underbrace{T^*_{U^*_k} \oplus \ldots \oplus T^*_{U^*_k}}_{n_k \text{ times}})$$

is essentially normal and $\operatorname{ind}(T - \lambda I) = n_k$ for $\lambda \varepsilon U_k$. The essential normality of T depends on the fact that $\operatorname{Area}(U_k) \to 0$.

11.6 Theorem. If T is an essentially normal operator on $\mathcal{H}$ such that $\operatorname{ind}(T-\lambda I) \leq 0$ for λ not in the essential spectrum of T, then there exists a subnormal operator S such that T is unitarily equivalent to some compact perturbation of S.

11.7 Corollary. If T is an essentially normal hyponormal operator on $\mathcal{H}$, then there exists a subnormal operator S such that T is unitarily equivalent to some compact perturbation of S.

Proof. If T is hyponormal, then $\ker(T-\lambda I) \subset \ker(T-\lambda I)^*$ and thus $\operatorname{ind}(T-\lambda I) \leq 0$ for λ not in the essential spectrum of T.

11.8 Theorem. If T is an essentially normal operator on $\mathcal{H}$, then there exist essentially normal operators T_1 and T_2 such that $\operatorname{ind}(T_1-\lambda I) \leq 0$ and $\operatorname{ind}(T_2-\lambda I) \geq 0$ for λ not in the essential spectrum of T_1 and T_2, respectively, and such that T is unitarily equivalent to some compact perturbation of $T_1 \oplus T_2$.

11.9 Corollary. If T is an essentially normal operator on $\mathcal{H}$, then there exist subnormal operators S_1 and S_2 such that T is unitarily equivalent to some compact perturbation of $S_1 \oplus S_2^*$.

Although there are problems associated with the notion of extension for other ideals of operators on a separable Hilbert space in higher dimensional cases (c.f. [25]), one can deal with the singly-generated case as follows. For $1 \leq p \leq \infty$, let $\mathcal{C}_p$ be the Schatten-von Neumann p-class, and denote by $\mathcal{E}_p(X)$ the set of unitary equivalence classes modulo $\mathcal{C}_p$ of operators T such that $[T,T^*] \in \mathcal{C}_p$ and $\sigma(\pi(T)) = X$. Then $\mathcal{E}_p(X)$ is a commutative semi-group relative to direct sum, and there is a natural homomorphism from $\mathcal{E}_p(X)$ to $\mathcal{E}_q(X)$ for $1 \leq p \leq q \leq \infty$.

For $p = \infty$, $\mathcal{E}_\infty(X) = \operatorname{Ext}(X)$, and $\mathcal{E}_1(X)$ is studied in [25]. Using the model described earlier, together with the trace estimate from [6] and an opposite estimate from [25], we see that the image of the map $\mathcal{E}_1(X) \to \mathcal{E}_\infty(X)$ can be described as follows. If $U_1, U_2, \ldots$ are the bounded components of $\mathbb{C}\setminus X$, then for given integers $n_1, n_2, \ldots,$ there exists $T \in \mathcal{E}_1(X)$ such that $\operatorname{ind}(T-\lambda I) = n_k$ for $\lambda \in U_k$ if and only if $\sum_k |n_k| \operatorname{Area}(U_k) < \infty$. For other values of p and q, the image of $\mathcal{E}_p(X) \to \mathcal{E}_q(X)$ has not been determined.

The map $\mathcal{E}_1(X) \to \mathcal{E}_\infty(X)$ is not generally injective. Helton and Howe [25] construct a homomorphism from $\mathcal{E}_1(X)$ to the space of real finite Borel measures on X; recent work of Carey and Pincus, using the theory of the determining function, shows that the image is $L^1(X,dxdy)$. This map is not injective, even for normal elements of $\mathcal{E}_1(X)$. (Note that any $T \in \mathcal{C}_2 \subset \mathcal{E}_1$ defines an element of the kernel.) In particular, the general problem of determining the unitary equivalence classes modulo $\mathcal{C}_p$ of normal operators with fixed essential spectrum is far from solved for $1 \leq p < \infty$. For example, it seems to be unknown whether two singular self-adjoint operators with the same essential spectrum are so related.

One way to sidestep the preceding difficulty is to define a notion of "stable equivalence" on $\mathcal{E}_p(X)$. Call two operators T_1 and T_2 with $[T_1,T_1^*],[T_2,T_2^*] \in \mathcal{C}_p$ "stably equivalent" in $\tilde{\mathcal{E}}_p(X)$ if there exist normal operators N_1 and N_2 with essential spectrum X such that $T_1 \oplus N_1$ is unitarily equivalent to some $\mathcal{C}_p$ perturbation of $T_2 \oplus N_2$. Again $\tilde{\mathcal{E}}_p(X)$ is a commutative semigroup, $\tilde{\mathcal{E}}_\infty(X) = \operatorname{Ext}(X)$, and there exists a homomorphism $\tilde{\mathcal{E}}_p(X) \to \tilde{\mathcal{E}}_q(X)$ for $1 \leq p \leq q \leq \infty$. Is the map $\tilde{\mathcal{E}}_p(X) \to \mathcal{E}_\infty(X)$ injective for $p > 1$? Is the induced homomorphism $\tilde{\mathcal{E}}_1(X) \to L^1(X,dxdy)$ injective? Further investigation of these semigroups seems to be of considerable interest.

The model described prior to Theorem 11.6 can be used to prove a recent result of Apostol, Foiaş, and Voiculescu [2] for essentially normal operators. (c.f. [18] for relevant definitions.)

<u>11.10 Theorem</u>. If T is a non-quasitriangular essentially normal operator, then there exists $\lambda \varepsilon \mathbb{C}$ such that $T-\lambda I$ is Fredholm and $\operatorname{ind}(T-\lambda I) < 0$.

<u>Proof</u>. If $\operatorname{ind}(T-\lambda I) \geq 0$ for all λ not in $\sigma(\pi(T))$,then T is unitarily equivalent to some compact perturbation of

$$N \oplus \sum_k \oplus (T^*_{U^*_k} \oplus \ldots \oplus T^*_{U^*_k}) \quad ,$$

where $U_1, U_2, \ldots$ are the bounded components of $\mathbb{C} \backslash \sigma(\pi(T))$, and $T^*_{U^*_k}$ is repeated $\operatorname{ind}(T-\lambda_k I)$ times, $\lambda_k \varepsilon U_k$. Since the eigenvalues of $T^*_{U^*_k}$ span the space, it follows easily (cf. [18], p.23) that $T^*_{U^*_k}$ is quasitriangular. Thus T is quasitriangular.

The converse is also valid (cf.[18]); namely, if T is an operator such that $T-\lambda I$ is Fredholm and $\operatorname{ind}(T-\lambda I) < 0$, then T is non-quasitriangular.

We now consider quasidiagonality [24]. Operators in $N+K$ are quasidiagonal, and if T is quasidiagonal, then both T and T* are quasitriangular. While the converse of the latter statement is not true in general, we have:

<u>11.11 Theorem</u>. If T is an essentially normal operator on H such that both T and T* are quasitriangular, then $T \varepsilon N+K$ and hence T is quasidiagonal.

<u>Proof</u>. If T and T* are quasitriangular, then $0 \leq \operatorname{ind}(T-\lambda I) = -\operatorname{ind}(T-\lambda I)^* \leq 0$ for λ not in the essential spectrum of T .
Thus $T \varepsilon N+K$ by Corollary 11.2.

<u>11.12 Corollary</u>. If T is a quasidiagonal essentially normal operator on H , then $T \varepsilon N+K$.

<u>11.13 Corollary</u>. If T is an essentially normal operator on H such that $p(T,T^*)$ is quasitriangular for each noncommutative

polynomial in T and T^*, then T is quasidiagonal.

It seems reasonable to ask whether this last corollary might not hold for general operators. Another important question is: Do all operators in $N+K$ have non-trivial invariant subspaces? By Corollary 11.2 an affirmative answer would yield the same result for essentially normal operators.

We conclude with a rather different question. Any *-automorphism $\theta: A \to A$ induces an automorphism $[\tau] \to [\theta\circ\tau]$ of $\mathrm{Ext}(X)$. If θ is inner, it was shown in Theorem 4.3 that the induced map is the identity, and it is natural to ask whether this is the case for arbitrary θ. If so, $\mathrm{Ext}(X)$ can be thought of as intrinsic with regard to A. Any θ restricts to an automorphism of the group G of invertible elements of A, and either $\mathrm{ind}\,\theta(g) = \mathrm{ind}(g)$ for all $g \in G$, or $\mathrm{ind}\,\theta(g) = -\mathrm{ind}(g)$ for $g \in G$. In the first case θ does induce the identity on $\mathrm{Ext}(X)$ if $X \subset \mathbb{C}$; we do not know whether the second case actually occurs. Note that any θ preserves the set of $a \in A$ with $\mathrm{ind}(a) = \infty$, and also the set for which $\mathrm{ind}(a) = -\infty$. Thus if θ reverses index, it would neither preserve quasitriangularity, nor send it to the dual property.

One way to attempt to get information about the automorphisms of A is to study the maximal abelian *-subalgebras (MASA's) of A. If these could be classified up to unitary equivalence, one would have some hold on the automorphism group modulo the inner automorphisms. Now [26] if M is a MASA in $L(H)$, then $\pi(M)$ is a MASA in A. These MASA's of A are easily classified, and all have totally disconnected maximal ideal space. (Note that strong and weak unitary equivalence does not agree for these MASA's.) By 5.8 there are MASA's of A without totally disconnected maximal ideal spaces. Since every MASA is the direct limit of its separable subalgebras, and since the theory of $\mathrm{Ext}(X)$ embraces all separable C*-subalgebras of A, $\mathrm{Ext}(X)$ may be useful in attacking this problem.

References

1. J. H. Anderson, On compact perturbations of operators, Canadian J. Math., to appear.

2. C. Apostol, C. Foiaş and D. Voiculescu, Some results on non-quastriangular operators IV, Rev. Roum. Math. Pures et Appl., 1973, to appear.

3. M. F. Atiyah, Global theory of elliptic operators, Proceedings of the International Conference on Functional Analysis and Related Topics, Tokyo, 1969, 21-30.

4. F. V. Atkinson, The normal solubility of linear equations in normed spaces, Mat. Sb. N.S. 28 (70) (1951), 3-14.

5. I. D. Berg, An extension of the Weyl-von Neumann Theorem to normal operators, Trans. Amer. Math. Soc. 160 (1971) 365-371.

6. C. A. Berger and B. I. Shaw, Self-commutators of multi-cyclic hyponormal operators are always trace class, Bull. Amer. Math. Soc. 79 (1973), to appear.

7. A. Brown and P. R. Halmos, Algebraic properties of Toeplitz operators, J. Reine Angew. Math. 213 (1964), 89-102.

8. L. G. Brown, Almost every proper isometry is a shift, Indiana J. Math., to appear.

9. L. G. Brown, R. G. Douglas and P. A. Fillmore, Extensions of C*-algebras, operators with compact self-commutators, and K-homology, Bull. Amer. Math. Soc. 79 (1973), to appear.

10. J. W. Calkin, Two-sided ideals and congruences in the ring of bounded operators in Hilbert space, Ann. Math. 42 (1941), 839-873.

11. L. A. Coburn, The C*-algebra generated by an isometry, Bull. Amer. Math. Soc. 73 (1967), 722-726.

12. L. A. Coburn, The C*-algebra generated by an isometry, II, Trans. Amer. Math. Soc. 137 (1969), 211-217.

13. L. A. Coburn, R. G. Douglas, D. G. Schaeffer and I. M. Singer, C*-algebras of operators on a half-space II, Index Theory, Publ. Math. I.H.E.S. 40 (1971), 69-79.

14. J. A. Deddens and J. G. Stampfli, On a question of Douglas and Fillmore, Bull. Amer. Math. Soc. 79 (1973), 327-330.

15. J. Dixmier, Les C*-algebres et Leurs Representations, Gauthier-Villars, Paris, 1964.

16. R. G. Douglas, Banach Algebra Techniques in the Theory of Toeplitz Operators, Lectures given at the CBMS Regional Conference at the University of Georgia, 1972.

17. R. G. Douglas, Banach Algebra Techniques in Operator Theory, Academic Press, New York, 1972.

18. R. G. Douglas and Carl Pearcy, Invariant subspaces of non-quasitriangular operators, these Notes.

19. S. Eilenberg and N. E. Steenrod, Foundations of Algebraic Topology, Princeton, 1952.

20. K. Ekman, Topological and analytic indices on C*-algebras, Dissertation, Mass. Inst. Tech., 1973.

21. P. A. Fillmore, J. G. Stampfli and J. P. Williams, On the essential numerical range, the essential spectrum, and a problem of Halmos, Acta Sci. Math. (Szeged) 33 (1972), 179-192.

22. P. R. Halmos, Permutations of sequences and the Schröder-Bernstein Theorem, Proc. Amer. Math. Soc. 19 (1968), 509-510.

23. P. R. Halmos, Continuous functions of Hermitian operators, Proc. Amer. Math. Soc. 81 (1972), 130-132.

24. P. R. Halmos, Ten problems in Hilbert space, Bull. Amer. Math. Soc. 76 (1970), 887-933.

25. J. W. Helton and R. E. Howe, Integral operators: commutators, traces, index, and homology, these Notes.

26. B. E. Johnson and S. K. Parrott, Operators commuting with a von Neumann algebra modulo the set of compact operators, J. Functional Anal. 11 (1972), 39-61.

27. J. S. Lancaster, Lifting from the Calkin algebra, Dissertation, Indiana University, 1972.

28. J. von Neumann, Charakterisierung des Spectrums eines Integraloperators, Hermann, Paris, 1935.

29. C. L. Olsen, A structure theorem for polynomially compact operators, Dissertation, Tulane University, 1970.

30. C. M. Pearcy and N. Salinas, Compact perturbations of semi-normal operators, Indiana Math. J. 22 (1973), 789-793.

31. C. M. Pearcy and N. Salinas, Operators with compact self-commutator, Can. J. Math., to appear.

32. C. A. Rickart, General Theory of Banach Algebras, Van Nostrand, Princeton, 1960.

33. J. G. Stampfli, Compact perturbations, normal eigenvalues, and a problem of Salinas, to appear.

34. F. Thayer-Fabrega, Liftings in the category of C*-algebras, Dissertation, Harvard University, 1972.

35. H. Weyl, Über beschrankte quadratischen Formen deren Differenz vollstetig ist, Rend. Circ. Mat. Palermo 27 (1909), 373-392.

S.U.N.Y. at Stony Brook
Stony Brook, N. Y.

Dalhousie University
Halifax, N. S.

$\mathcal{E}xt(X)$ FROM A HOMOLOGICAL POINT OF VIEW

Jerome Kaminker and Claude Schochet[1]

I. Introduction

In a series of recent papers [BDF 1,2,3], L.G. Brown, R.G. Douglas, and P.A. Fillmore (abbreviated BDF) study a certain class of extensions of C*-algebras, i.e., short exact sequences of the form

$$0 \to K(H) \to E \to C(X) \to 0$$

where X is compact metric, $K(H)$ is the ideal of compact operators, and E is a C*-algebra containing $K(H)$ and the identity, and contained in $L(H)$. $\mathcal{E}xt(X)$ is defined to be the set of unitary equivalence classes of such extensions. In [BDF 1] it is shown that $\mathcal{E}xt(X)$ is a commutative semigroup with identity. In [BDF 3] it is shown that for appropriate X, $\mathcal{E}xt(X)$ is equal to homology K-theory $K_1(X)$. Explicit computations of $\mathcal{E}xt(X)$ for X a subset of the plane yield important results in operator theory.

This note is devoted to describing $\mathcal{E}xt(X)$ in terms more analogous to those of homological algebra. Specifically, it is shown that $\mathcal{E}xt(X) = \mathrm{Ext}^1(C(X), K(H))$, a functor which is defined via essentially standard techniques as the right derived functor of $\hom^u(C(X), K(H))$. This description of $\mathcal{E}xt(X)$ is quite suggestive and amenable to generalization. We hope to pursue these matters subsequently.

2. C* H-spaces and Unitary Equivalence

Fix notation as follows:

H = an infinite dimensional separable complex Hilbert space

$L(H)$ = the bounded operators on H

[1] We wish to thank John Conway, Joe Stampfli and others at Indiana University for many helpful conversations.

$K(H)$ = the ideal of compact operators in $L(H)$. It is the unique maximal closed two-sided ideal in $L(H)$.

$A(H) = L(H)/K(H)$, the Calkin algebra

$U(B)$ = the unitary operators in the C*-algebra B

X = a compact, separable, metric space

Suppose $u: H_1 \to H_2$ is an isomorphism of Hilbert spaces. (Isomorphisms are onto by definition). Then u induces a C*-isomorphism $L(u) : L(H_1) \to L(H_2)$ defined by $L(u)T = uTu^*$ which preserves the compacts and hence induces a C*-isomorphism $A(u): A(H_1) \to A(H_2)$. Note that $L(wu) = L(w)L(u)$ and similarly for A.

Fix an isomorphism $s: H \oplus H \to H$. (If s' is a different choice then $us = s'$ for some isomorphism $u: H \to H$.) The natural injections $\iota_1, \iota_2 : H \to H \oplus H$ induce C*-injections after applying K, L, or A, and hence a commutative diagram of C*-morphisms

$$\begin{array}{ccccc} K(H) \oplus K(H) & \to & K(H \oplus H) & \xrightarrow{} & K(H) \\ \downarrow & & \downarrow & L(s) & \downarrow \\ L(H) \oplus L(H) & \to & L(H \oplus H) & \to & L(H) \\ \downarrow & & \downarrow & & \downarrow \\ A(H) \oplus A(H) & \to & A(H \oplus H) & \to & A(H) \end{array}$$

<u>Definition 2.1</u>. Let

$$m_K : K(H) \oplus K(H) \to K(H)$$

$$m_L : L(H) \oplus L(H) \to L(H)$$

$$m_A : A(H) \oplus A(H) \to A(H)$$

be the three horizontal composites in diagram 2.1. These are external multiplications on $K(H)$, $L(H)$, $A(H)$ respectively.

<u>Definition 2.2</u>. Let $B = K(H)$, $L(H)$, or $A(H)$ and let A be some C*-algebra. Define <u>$\hom^u(A,B)$</u> to be the set of C*-morphisms from A to B modulo the following equivalence relation-- called "<u>unitary equivalence</u>" and written "$\sim$" :

a) $B = K(H)$; $f,g : A \to B$ are equivalent if there exists $T \in U(L(H))$ with $T^*(fa)T = ga$ for every $a \in A$.

b) $B = L(H)$ or $A(H)$; $f,g : A \to B$ are equivalent if there exists $T \in U(B)$ with $T^*(fa)T = ga$ for every a in A .

Remark 2.3. The morphisms m_K, m_L, m_A are unique up to unitary equivalence. Definition 2.2 is clearly ad hoc, and there should be a general notation of equivalence to cover both cases.

Definition 2.4. A C* H-space (B,μ) is a C*-algebra B together with a C*-injection $\mu : B \oplus B \to B$ such that $\mu(\mu \times 1) \sim \mu(1 \times \mu)$ and $\mu t \sim \mu$, where $t(x,y) = (y,x)$.

Remark 2.4. We chose the name C* H-space by analogy with the notion of H-space in topology. There, an operation is defined on a space, but it possesses properties such as associativity or commutativity only up to homotopy. Here we have commutativity and associativity only up to unitary equivalence. Remarkably enough, homotopy and unitary equivalence, in this context, are not entirely unrelated. (See Remark 4.7 at end of paper).

As the definition stands, B must be a C*-algebra for which $\hom^{U}(A,B)$ makes sense. Evident samples of C* H-spaces and morphisms are

$$(K(H),\, m_K) \to (L(H),\, m_L) \to (A(H),\, m_A)$$

Proposition 2.5. Let (B,μ) be a C* H-space. Then $\hom^{U}(A,B)$ is a commutative semigroup.

Proof: This is immediate from the definition of the operation $\oplus$ which is the dotted arrow below:

$$\begin{array}{ccc} \text{hom}(A,B) \times \text{hom}(A,B) = \text{hom}(A,B\oplus B) & \xrightarrow{\mu_*} & \text{hom}(A,B) \\ \downarrow & & \downarrow \\ \text{hom}^u(A,B) \times \text{hom}^u(A,B) & \dashrightarrow & \text{hom}^u(A,B) \end{array}$$

Corollary 2.6. The natural maps

$$\text{hom}^u(A, K(H)) \xrightarrow{i_*} \text{hom}^u(A, L(H)) \xrightarrow{\pi_*} \text{hom}^u(A, A(H))$$

are morphisms of commutative semigroups. Further, i_* is injective and $\pi_* i_* = 0$.

III. Extensions of C*-algebras

In this section we recall the definition of $Ext(X)$ as in [BDF, 1]. Let A and C be C*-algebras, with C having an identity and $A \subset L(H)$. An extension of A by C is a commutative diagram

$$\begin{array}{ccccccccc} & & & \nearrow & L(H) & & & & \\ & & & & \uparrow & \phi & & & \\ 0 & \to & A & \to & B & \to & C & \to & 0 \end{array}$$

where B is a C*-algebra with identity, ϕ is a C*-morphism onto C, and $\ker(\phi) = A$. It is denoted (B,ϕ). An extension of the form

$$0 \to K(H) \to E \xrightarrow{\phi} C \to 0$$

induces a canonical C*-injection τ_ϕ, defined to be the composite

$$C \xrightarrow{\phi^{-1}} E/K(H) \to L(H)/K(H) = A(H) .$$

Conversely, a C*-injection $\tau: C \to A(H)$ determines an extension (E_τ, ϕ_τ) of $K(H)$ by C, taking it to be the pullback of the diagram

$$\begin{array}{ccccccccc} 0 & \to & K(H) & \to & L(H) & \to & A(H) & \to & 0 \\ & & \| & & \uparrow & & \uparrow \tau & & \\ 0 & \to & K(H) & \to & E_\tau & \to & C & \to & 0 \end{array}$$

One wishes to classify extensions up to unitary equivalence, as follows:

<u>Definition 3.1</u>. Two extensions (E_1,τ_1) , (E_2,τ_2) of $K(H_i)$ by $C(X)$ are (unitarily) <u>equivalent</u> if there exists a C*-isomorphism $\psi : E_1 \to E_2$ with $\phi_2\psi = \phi_1$. Define $Ext(X)$ to be equivalence classes of extensions of $K(H)$ by $C(X)$.

Let $inj(A,B)$ be the subset of $hom(A,B)$ consisting of the injections $A \to B$. If f is an injection and $f \sim g$ (unitary equivalence) then g is an injection. So $inj^u(A,B)$ is defined (as a subset of $hom^u(A,B)$.) Let $\theta_X : Ext(X) \to inj^u(C(X), A(H))$ by $\theta_X(E,\phi) = \tau_\phi$. Then it is easy to check

<u>Proposition 3.2</u>. For each X , $\theta_X : Ext(X) \to inj^u(C(X), A(H))$ is an isomorphism of sets.

Following BDF, recall that $Ext(X)$ is a commutative semigroup with identity τ_X via a modified Baer sum construction. Prop. 2.5 gives $hom^u(C(X), A(H))$ a commutative semigroup structure and $inj^u(C(X), A(H))$ inherits the same structure. So it makes sense to claim

<u>Proposition 3.3</u>. For each X , $\theta_X : Ext(X) \to inj^u(C(X), A(H))$ is an isomorphism of commutative semigroups.

<u>Proof</u>: Let (E_i,ϕ_i) be an extension over $C(X)$ with $E_i \subset L(H_i)$ and $\theta_X(E_i,\phi_i) = \tau_i$, for $i = 1,2$. Let $(E,\phi) = (E_1,\phi_1) \oplus (E_2,\phi_2)$. Let $r_i : H_i \to H$ be isomorphisms and define r_3 to be the composite

$$H_1 \oplus H_2 \xrightarrow{r_1 \oplus r_2} H \oplus H \xrightarrow{s} H$$

where s is some fixed isomorphism. (The composite is unique up to unitary equivalence.) Then $\tau_1 \oplus \tau_2$ is the composite

$$C(X) \xrightarrow{(\tau_1,\tau_2)} A(H_1) \oplus A(H_2) \xrightarrow{A(r_3)i} A(H) .$$

On the other hand, $\theta_X(E,\phi) = \tau$, and τ is characterized by the classifying diagram

$$\begin{array}{ccccccccc} 0 & \to & K(H_1 \oplus H_2) & \to & E & \xrightarrow{\phi} & C(X) & \to & 0 \\ & & \| & & \downarrow & & \downarrow \tau & & \\ 0 & \to & K(H_1 \oplus H_2) & \to & L(H_1 \oplus H_2) & \xrightarrow{\pi} & A(H_1 \oplus H_2) & \to & 0 \end{array}$$

But the diagram

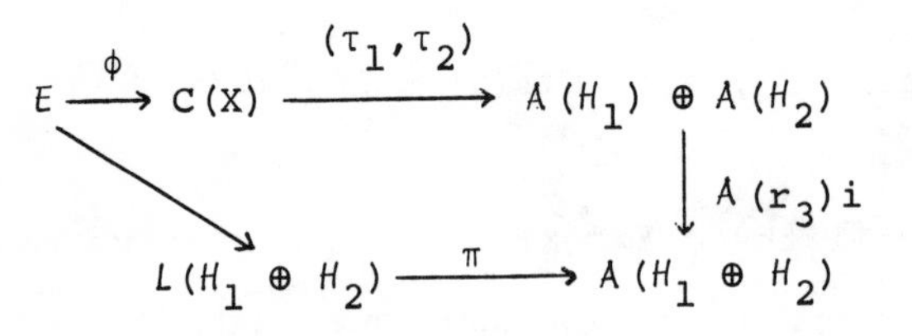

commutes, so we may choose τ to be the composite $A(r_3)i(\tau_1,\tau_2) = \tau_1 \oplus \tau_2$.

<u>Definition 3.4.</u> A C*-algebra extension $0 \to A \to B \xrightarrow{\phi} C \to 0$ is <u>split</u> if there exists a C*-morphism $\sigma : C \to B$ such that $\phi\sigma = id_C$.

For extensions of $K(H)$ by $C(X)$ classified by $\tau : C(X) \to A(H)$ a splitting corresponds to a C*-injection $\sigma : C(X) \to L(H)$ lifting τ ; i.e., $\pi\sigma = \tau$.

<u>Proposition 3.5.</u> [BDF 1] A split extension of $K(H)$ by $C(X)$ exists. Any two split extension are equivalent, so there is a unique class $\tau_X \varepsilon Ext(X)$. This class is the identity for the semigroup $Ext(X)$.

For semigroups S,T with identities $1_S, 1_T$ and a morphism $f : S \to T$, the kernel of f is defined by $\ker(f) = f^{-1}(1_T)$.

Proposition 3.6. The sequence

$$0 \to \hom^{u}(C(X),K(H)) \xrightarrow{i_*} \hom^{u}(C(X),L(H)) \xrightarrow{\pi_*} \hom^{u}(C(X),A(H))$$

is exact as semigroups. That is, i_* is injective and $\mathrm{im}(i_*)=\ker(\pi_*)$.

Proof: Only $\ker(\pi_*) \subset \mathrm{im}(i_*)$ requires an argument. It follows since $\ker(\pi) = \mathrm{im}(i)$.

4. *Ext* as a derived functor

Let $\underline{M}$ be the category of C* H-spaces and let $\underline{A}$ be the category of commutative semigroups. Morphisms preserve identities when present. Proposition 3.6 implies:

Theorem 4.1. The covariant functor $\hom^{u}(C(X), \text{——})$ from $\underline{M}$ to $\underline{A}$ is left exact. In particular, if $0 \to A \to B \to C \to 0$ is exact in $\underline{M}$, then

$$0 \to \hom^{u}(C(X),A) \to \hom^{u}(C(X),B) \to \hom^{u}(C(X),C)$$

is exact in $\underline{A}$.

Recall that if T is a sub-semigroup of the commutative semigroup S (sharing a common identity) then the quotient S/T is defined to be S modulo the equivalence relation defined by $s_1 \sim s_2$ if $s_1 + t_1 = s_2 + t_2$ for $s_i \in S$, $t_i \in T$. If $f : S' \to S$ is a morphism of commutative semigroups with identity, then the cokernel of f is defined by $\mathrm{coker}(f) = S/\mathrm{im}(f)$.

Theorem 4.2. Let X be compact, separable, metric. Then

$$Ext(X) \cong \mathrm{coker}(\pi_*) = \hom^{u}(C(X), A(H)) \Big/ \pi_*\hom^{u}(C(X),L(H))$$

as commutative semigroups with identity.

Proof: In light of Prop. 3.3, it suffices to exhibit an isomorphism of semigroups

$$\Phi_X : \mathrm{coker}(\pi_*) \to \mathrm{inj}^u(C(X), A(H))$$

If $\tau : C(X) \to A(H)$, denote its unitary equivalence class by $\bar{\tau}$. Let $\tau_X : C(X) \to A(H)$ represent the trivial (split) extension. Note that if $\tau : C(X) \to A(H)$ is any C*-morphism, then $\tau \oplus \tau_X$ is a C*-injection. Define

$$\phi_X : \mathrm{hom}^u(C(X), A(H)) \to \mathrm{inj}^u(C(X), A(H))$$

by $\phi_X(\bar{\tau}) = \overline{\tau \oplus \tau_X}$. Note that $\overline{\tau_1 \oplus \tau_2} = \bar{\tau}_1 \oplus \bar{\tau}_2$. Now ϕ_X is a semigroup morphism, for

$$\begin{aligned} \phi_X(\bar{\tau}_1 \oplus \bar{\tau}_2) &= \phi_X(\overline{\tau_1 \oplus \tau_2}) \\ &= \overline{\tau_1 \oplus \tau_2 \oplus \tau_X} \\ &= \overline{\tau_1 \oplus \tau_X} \oplus \overline{\tau_2 \oplus \tau_X} \\ &= \phi_X(\bar{\tau}_1) \oplus \phi_X(\bar{\tau}_2) \end{aligned}$$

Furthermore, ϕ_X is onto, since if τ is an injection, then

$$\bar{\tau} = \overline{\tau \oplus \tau_X} = \phi_X(\bar{\tau}) .$$

To complete the proof it suffices to show that ϕ_X respects the equivalence relation (hence inducing the map Φ_X) and that Φ_X is injective.

Recall that by definition, the identity τ_X lifts; $\bar{\tau}_X = \pi_*(\bar{\sigma}_X)$, where σ_X is injective. Then, writing $\bar{\tau}_1 \oplus \pi_*\bar{\sigma}_1 = \bar{\tau}_2 \oplus \pi_*\bar{\sigma}_2$,

$$\begin{aligned} (\bar{\tau}_1 \oplus \pi_*\bar{\sigma}_1) \oplus \bar{\tau}_X &= \bar{\tau}_1 \oplus \pi_*(\bar{\sigma}_1 \oplus \bar{\sigma}_X) \\ &= \bar{\tau}_1 \oplus \bar{\tau}_X = \phi_X(\bar{\tau}_1) . \end{aligned}$$

Similarly, $(\bar{\tau}_2 \oplus \pi_*\bar{\sigma}_2) \oplus \bar{\tau}_X = \phi_X(\bar{\tau}_2)$ and hence $\phi_X(\bar{\tau}_1) = \phi_X(\bar{\tau}_2)$. So Φ_X is defined via the diagram

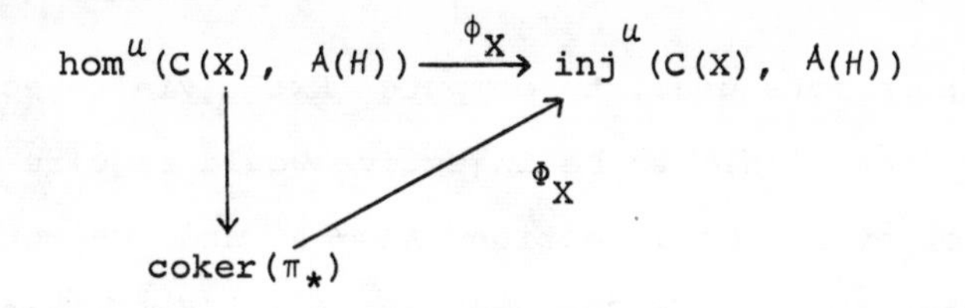

Finally, if $\Phi_X(\bar{\tau}_1) = \Phi_X(\bar{\tau}_2)$, then $\bar{\tau}_1 \oplus \bar{\tau}_X = \bar{\tau}_2 \oplus \bar{\tau}_X$ and since $\bar{\tau}_X \in \operatorname{im}(\pi_*)$, we have $\bar{\tau}_1 = \bar{\tau}_2$ in coker (π_*) .

A more categorical expression for $Ext(X)$ is at hand. The functor $\hom^u(C(X), \text{——}) : \underline{M} \to \underline{A}$ is left exact, so with appropriate generalization of homological techniques [G] , it has right derived functors. These are traditionally labelled $\operatorname{Ext}^n(C(X), \text{——})$. In particular, $\operatorname{Ext}^1(C(X), K(H))$ may be computed as $\operatorname{Ext}^1(C(X), K(H)) = \operatorname{coker}(\pi_*)$, where

$$0 \to K(H) \to I \to I/K(H) \to 0$$

is an acyclic presentation of $K(H)$ and $\pi_* : \hom^u(C(X), I) \to \hom^u(C(X), I/K(H))$. Fortunately, the classifying extension

$$0 \to K(H) \to L(H) \to A(H) \to 0$$

is an acyclic presentation of $K(H)$. (The acyclicity of $L(H)$ follows since there are only trivial extensions of $L(H)$ by $C(X)$. X must be empty.) Plugging in definitions, we obtain a restatement of Theorem 4.2. :

<u>Theorem 4.3</u>. $Ext(X) = \operatorname{Ext}^1(C(X), K(H))$ as commutative semigroups with identity.

<u>Remark 4.4</u>. This brings us full circle back to the simplest case of extensions. If A and C are abelian groups, then extensions of A

by C are classified by $\mathrm{Ext}^1(C,A)$. Similar results hold for extensions of other types of algebraic objects.

<u>Remark 4.5</u>. It is more usual to compute Ext^1 via injective presentations. For $L(H)$ to be injective would require the following to be true: for every C*-injection $A' \to A$ and C*-morphism $A' \to L(H)$, then there is a C*-morphism $A \to L(H)$ making

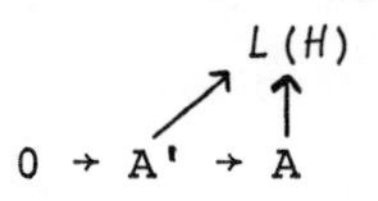

commute. This is true when A' is a closed two-sided ideal of A, but is false in general. It would be even more usual to compute Ext^1 via a projective resolution of $C(X)$. It is not clear that such resolutions exist, but the fact that $\mathcal{E}xt$(Hilbert cube) = 0 suggests that C(Hilbert cube) may be close to projective.

<u>Remark 4.6.</u> Our Theorem 4.3. is the analog of a theorem of Busby [B]. His result would classify extensions of the form $0 \to K(H) \to E \to C(X) \to$ (under an equivalence relation stronger than unitary equivalence) by $\mathrm{hom}(C(X), \mathcal{O}(K(H)))$, where $\mathcal{O}(K(H))$ is the quotient of $M(K(H))$, the algebra of double centralizers of $K(H)$, by $K(H)$. Since $M(K(H)) = L(H)$, one gets $\mathcal{O}(K(H)) = A(H)$. If one considers extensions with $K(H)$ replaced by a Schatten class, C_p, then $M(C_p)$ is still all the bounded operators, so that $\mathcal{O}(C_p) = L(H)/C_p$. One might hope that $\mathrm{Ext}^1(CX, \mathcal{O}(C_p))$ would classify extensions in this case as well.

<u>Remark 4.7</u>. We conclude by stating a question which has arisen in our study and seems to be of general interest.

Question: Does the natural projection $\pi : L(H) \to A(H)$ have the C* covering homotopy property for maps from $C(X)$, in the sense that a map $\hat{H}$ exists completing the following diagram

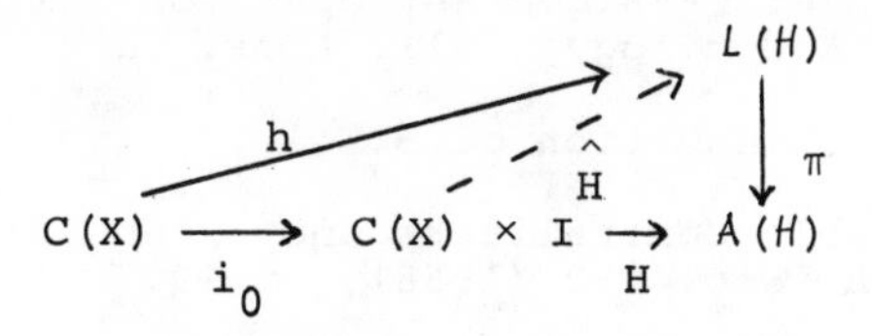

where h is a C* morphism and $H(.,t)$, $\hat{H}(.,t)$ are both C* morphisms for each t ?

Partial results are known to several people. In particular John Conway has proved that the C* covering homotopy property holds when X is totally disconnected.

If we now define $\text{inj}^h(CX, A(H))$ to be homotopy classes of C* injections (through C* injections) and use the fact that $Ext(X)$ is actually a group, [BDF 2], then we can show

$$Ext(X) = \text{inj}^u(C(X), A(H)) = \text{inj}^h(C(X), A(H))$$

whenever X is a space for which the C* covering homotopy property holds. It is interesting to observe that $\text{inj}^h(C(X), A(H))$ is a homotopy invariant. A positive answer to this question would provide an alternative proof of the homotopy invariance of $Ext(X)$ (announced in [BDF 2]) and shed light on the relationship between homotopy and unitary equivalence.

REFERENCES

[BDF 1] L. G. Brown, R. G. Douglas, and P. A. Fillmore, Unitary equivalence modulo the compact operators and extensions of C*-algebras, these Notes.

[BDF 2] ———, Extensions of C*-algebras, operators with compact self-commutators, and K-homology, Bull. Amer. Math. Soc. (to appear).

[BDF 3] ———, untitled continuation of BDF 1 .

[B] Robert C. Busby, Double Centralizers and extensions of C*-algebras, Trans. Amer. Math. Soc. 132 (1968), 79-99.

[G] A. Grothendieck, Technique de descente et théorèmes d'existence en géométrie algébrique I, II, Sem. Bourbaki 190, 195(1959/60).

Indiana University
Bloomington, Indiana

INTEGRAL OPERATORS: TRACES, INDEX, AND HOMOLOGY

J. William Helton and Roger E. Howe

This paper begins an investigation of certain concrete phenomena in abstract operator algebras. The main examples of the algebras we study are algebras of singular integral operators (pseudodifferential operators of order zero). As everyone knows, the Atiyah-Singer Index Theorem gives an elegant formula for computing the index of a pseudodifferential operator in terms of its symbol. What we are doing might be thought of analogously. We are attempting to develop a theory of traces of commutators and higher commutators which gives explicit but elegant formulas for their computation. As we shall see, the trace theory and index theory are closely related, and it is conceivable that full development of the trace theory could lead to an abstract and generalized version of the Index Theorem. However, at the moment this goal is far off.

One function of this trace theory is that it serves as a link between several existing areas of work. It is closely related to two articles appearing in this volume, namely, to the paper on extensions of C*-algebras by Brown, Douglas, and Fillmore, and to the one on trace norm estimates by Berger and Shaw. A determinant theory parallel to the trace theory exists (at least in dimension one). This is begun in Part I, §10, and has been pursued by L. Brown in an article which also appears in this volume. Our work is very closely related to that of J.D. Pincus, and our initial interest in this subject was motivated in part by a desire to understand his work (see [P_1], [P_2], [P_4]).

The present paper is divided into two parts. Part I treats

This work was partially supported by the N.S.F.

abstract one-dimensional singular integral operators and is purely operator-theoretic. It considers an operator algebra generated by one operator and studies traces of commutators in that algebra. The results in Part I preceded our general study; we have had them for some time and our program in this case is well-developed. In Part II we describe the results of Part I in terms of homology theory. This interpretation suggests extension of these results to algebras having many generators and to algebras of higher-dimensional singular integral operators. (Beyond foundational results, however, we can deal only with the one- and two-dimensional cases.) We confine ourselves in this paper to announcing existing results and to describing the general program. A detailed presentation of this work requires considerable borrowing from algebraic topology and will be carried out elsewhere.

The paper is overly long in that after the proof of the main theorem is complete we spend much time in describing the many directions one can take. A casual reader interested only in a survey of results is advised to read the introduction to Part I and then §1,2, and 3 of Part II before proceeding to the body of Part I. The reader interested in seeing the proof of the main theorem may take courage from the fact that all of the crucial ideas are in Part I, §§2 - 6. The authors would like to thank their colleagues Larry Brown, Ron Douglas, and Joel Pincus for interesting discussions.

TABLE OF CONTENTS

PART I. Almost commuting pairs of operators

0. Introduction
1. An example: Toeplitz operators
2. Wallach's Lemma
3. Functional Calculus and extension of (,)
4. The trace form and the essential spectrum
5. The Fredholm index
6. Putting together the pieces
7. The transformation law
8. Smoothness of $g(\mu,\nu)$; connections with other work
9. Effect of compact perturbations; relation with Brown-Douglas-Fillmore.
10. A determinant theory

Appendix: On the proof that the trace invariant is a measure.

PART II. Algebras of higher dimensional singular integral operators

0. Introduction
1. A different view of almost commuting pairs
2. Higher dimensional singular integral operators and higher trace forms
3. Index Theory
4. Lower homology classes
5. C^{∞}-extensions

Part I. ALMOST COMMUTING PAIRS OF OPERATORS

An almost commuting (a.c.) pair of operators on a Hilbert space H is a pair X,Y of bounded self-adjoint operators with trace class commutator $[X,Y] = XY - YX$. If X,Y are an a.c. pair, then $T = X + iY$ has self-commutator $[T^*,T] = 2i[X,Y]$ which is trace class, and conversely. Thus, the study of a.c. pairs is equivalent to the study of operators T with $[T^*,T]$ trace class. However, it is useful to think of X and Y as independent variables. Since every trace class operator is compact this situation is a special case of the one studied by Brown-Douglas-Fillmore. A strong point of the trace class situation is that one can compute many things quite explicitly and that it arises often.

Almost commuting pairs come up in several branches of operator theory. The two main concrete classes of examples of almost commuting pairs of operators are Toeplitz (or Wiener-Hopf) operators with smooth symbol and singular integral operators on the line. The singular integral operators actually provide essentially all examples since there is a representation theorem [P_1], [R], [X] stating that an a.c. pair X,Y may be simultaneously modelled as multiplication by x and a singular integral operator on a suitable L^2-space of vector-valued functions on $\mathbb{R}$. In other words in the spectral representation for X the operator Y becomes a singular integral operator. Thus, in a sense the study of a.c. pairs is an abstract study of algebras generated by multiplication and singular integral operators. The Toeplitz operators give a nice example of our main results and we will begin in §1 with explicit computations for the Toeplitz case.

Along more abstract lines, there is considerable overlap of the theory of a.c. pairs with the theory of hyponormal operators. (Recall T is hyponormal if $[T^*,T] \geq 0$.) This is due to theorems of Kato [K] and more recently of Putnam [Put_1] and Berger-Shaw [B-S] which say that hyponormal operators satisfying various finiteness conditions have trace class self-commutator. We note also that the basic examples from the Toeplitz and singular integral operator classes are hyponormal as well as almost commuting. As we will point out, various of our results may be strengthened or simplified if we assume for our a.c. pair X,Y that $2i[X,Y] \geq 0$, i.e. that $T = X+iY$ is hyponormal. A good example of this is the result of Clancey and of Pincus $[P_4]$, for which we give a new proof in §8.

One of the basic results on hyponormal operators is Putnam's inequality $[Put_1]$, [C], [H], which says for T hyponormal $||[T^*,T]|| \leq \frac{1}{\pi}\mu(\sigma(T))$ where $||\ \ ||$ denotes operator norm, μ is Lebesgue measure and $\sigma(T)$ is the spectrum of T. The work of Pincus suggested that the relevant quantity was not $||[T^*,T]||$ but $tr([T^*,T])$, where tr denotes trace. (Since $[T^*,T] \geq 0$, we take $tr([T^*,T]) = +\infty$ if $[T^*,T]$ is not trace class.) Recently, inequalities involving $tr([T^*,T])$ have been appearing $[Put_1]$, [B-S]. We feel that our main theorem provides a context in which to set these results. See §8 for more details. We also feel that the interpretative discussion in Part II sheds light on the role played by $tr([T^*,T])$ and on the significance of Lebesgue measure in these inequalities.

We begin the detailed part of the introduction by recalling some basic definitions. Let $L(H)$ be the algebra of all bounded operators on H and $LC(H)$ be the closed ideal in $L(H)$ of all compact operators. Then $L(H)/LC(H)$ is called the <u>Calkin algebra</u> and $e\sigma(T)$, the <u>essential spectrum</u> of T, is the spectrum of the image of T in this algebra. The C*-algebra $\mathfrak{A}(T)$ generated by T is the same as the C*-algebra $\mathfrak{A}(X,Y)$ generated by X and Y, and $e\sigma(T)$ is just the maximal ideal space of the commutative C*-algebra $\mathfrak{A}(T)/\mathfrak{A}(T)\cap LC(H)$. This algebra under the Gelfand representation is isometrically isomorphic to the continuous functions $C(e\sigma(T))$ on $e\sigma(T)$. The set $e\sigma(T)$ under the natural identification of $\mathbb{C}$ with

$\mathbb{R}^2$ is the <u>joint essential spectrum</u> of X and Y . We shall be considering a skew symmetric bilinear form $\langle\ ,\ \rangle$ on a vector space $\mathfrak{A}$. The <u>radical</u> of $\langle\ ,\ \rangle$ is $\{b; \langle b,a\rangle = 0 \text{ if } a \in \mathfrak{A}\}$.

For the moment let us be fanciful. We wish to study an a.c. pair X,Y of operators. Suppose that one could define a simple, intrinsically interesting bilinear form $\langle\ ,\ \rangle$ on $\mathfrak{A}(X,Y)$ with the property that $LC \cap \mathfrak{A}(X,Y)$ belonged to its radical. Then $\langle\ ,\ \rangle$ would induce a form on $\mathfrak{A}(X,Y)/LC \cap \mathfrak{A}(X,Y)$ which we could simply think of as a bilinear form on the space $C(e\sigma(T))$. If one could classify and represent such forms, then the representation would supply intrinsic information about X and Y . Our basic approach is very much like this. The bilinear form we would like to study is $\langle A,B\rangle = tr([A,B])$, however, it is defined only on a dense subalgebra of $\mathfrak{A}(X,Y)$. Thus, the program we just described must be modified. The modifications turn out to be quite interesting in themselves since one is forced to develop a 'C^∞' rather than a 'continuous' theory.

Now we give a precise description of our main theorem. Let X and Y be an a.c. pair of self-adjoint operators Let $P\mathfrak{A}(X,Y)$ be the operator algebra of formal polynomials in X and Y . A simple calculation shows that the commutator ideal (the ideal generated by all commutators) of $P\mathfrak{A}(X,Y)$ consists of trace class operators. Thus, the bilinear form $\langle R,S\rangle = tr([R,S])$ with $R,S \in P\mathfrak{A}$ is defined on $P\mathfrak{A}$. If H were finite dimensional $\langle\ ,\ \rangle$ would be identically zero; in infinite dimensions it certainly need not be. However, if either R or S is trace class then $\langle R,S\rangle = 0$. Therefore, the commutator ideal I of $P\mathfrak{A}$ is in the radical of $\langle\ ,\ \rangle$, and it induces a form $(,)$ on the commutative algebra $P\mathfrak{A}/I$, a quotient algebra of the polynomials on $\mathbb{R}^2$.

The most concrete way of viewing the induced form is as follows. Let P_2 denote the polynomials on $\mathbb{R}^2$; it is a commutative algebra of functions. If $p(x,y) = \Sigma\, a_{nm} x^n y^m$ is in P_2 , then define the operator $p(X,Y)$ in $\mathcal{PO}$ to be $\Sigma\, a_{nm} X^n Y^m$; here powers of the operator X precede powers of the operator Y . Define for p,q in P_2

$$(p,q) = \langle p(X,Y),\ q(X,Y)\rangle = \mathrm{tr}[p(X,Y),\ q(X,Y)]$$

Note that if $\tilde{p}(X,Y)$ is an operator which arises from p by substituting X and Y into $\Sigma\, a_{nm} X^n Y^m$ in a different order, then $\tilde{p}(X,Y) - p(X,Y)$ is in I and so $(p,q) = \langle\tilde{p}(X,Y)\ ,\ \tilde{q}(X,Y)\rangle$. This work is devoted to a study of $(\ ,\ )$.

If $R,S \in \mathcal{PO}$ commute then it is clear that $\langle R,S\rangle = 0$. One occasion when R and S will certainly commute is when they are both polynomials of a third element of $\mathcal{PO}$. Translating this fact to P_2, we see that $(\ ,\ )$ has the property that it vanishes identically when restricted to any singly generated subalgebra. This property is crucial. We name a bilinear form on P_2 with this property a <u>Pincus form</u> in honour of J.D. Pincus and the property itself we call the <u>collapsing</u> property.

There is a very pretty representation for Pincus forms. We would like to thank Nolan Wallach for the elegant proof he supplied us of this representation. Our original proof was horrible; Wallach's argument, which we present in §2 , clearly goes to the heart of the matter.

<u>Wallach's lemma</u>: If $(\ ,\ )$ is a Pincus form on P_2 , then there is an unique linear functional ℓ on P_2 such that for $p,q \in P_2$

$$(p,q) = \ell\left(\frac{\partial p}{\partial x}\frac{\partial q}{\partial y} - \frac{\partial q}{\partial x}\frac{\partial p}{\partial y}\right) .$$

We note that $\frac{\partial p}{\partial x}\ \frac{\partial q}{\partial y} - \frac{\partial q}{\partial x}\frac{\partial p}{\partial y}$ is just the Jacobian of the mapping from $\mathbb{R}^2$ to $\mathbb{R}^2$ defined by p and q . Hence we write it as $J(p,q)$. Sometimes $J(p,q)$ is also called the Poisson bracket of p and q .

Initial estimates on the continuity of the form $\langle\ ,\ \rangle$ show that $(,)$ extends from $P_2(x,y)$ to a separately C^3 continuous function on $S(\mathbb{R}^2)$, so that the form ℓ attached to $(,)$ by Wallach's lemma is a distribution of compact support on $\mathbb{R}^2$. Here $S(\mathbb{R}^2)$ denotes the Schwartz space of all rapidly decreasing smooth functions. Further detailed study of $< , >$ leads to our main result.

Theorem: Let X,Y be an a.c. pair. Then the bilinear form $< , >$, initially defined on the algebra $P\mathfrak{A}(X,Y)$ of non-commutative polynomials in X and Y by the formula $<R,S> = \mathrm{tr}([R,S])$ for $R,S \in P\mathfrak{A}$, gives rise to a form $(,)$ on $S(\mathbb{R}^2)$ as described above. The form $(,)$ has the representation

$$(1) \qquad (f,g) = \int J(f,g)\,dP \quad \text{for} \quad f,g \in C^\infty(\mathbb{R}^2)$$

Thus for polynomials $p,q \in P_2$ we have

$$(1') \qquad \mathrm{tr}([p(X,Y),q(X,Y)]) = \int J(p(x,y),q(x,y))\,dP .$$

Here, $J(f,g)$ is the Jacobian of f and g and dP is a measure of compact support. The measure dP has in addition the following properties.

Identify $\mathbb{R}^2$ with $\mathbb{C}$ by $(x,y) \leftrightarrow x + iy$ and put $T = X + iY$. Then

i) the measure dP is supported on $\sigma(T)$, the spectrum of T.

ii) The measure dP is a constant multiple of Lebesgue measure on each connected component of the complement of the essential spectrum $e\sigma(T)$ of T.

iii) If $U \subseteq \mathbb{C}$ is a component of $\mathbb{C} \sim e\sigma(T)$, and if $\lambda \in U$, then on U,

$$\frac{dP}{d\mu} = \frac{-1}{2\pi i} \; \mathrm{ind}\,(T-\lambda) ,$$

where $d\mu$ is Lebesgue measure and $\mathrm{ind}(T-\lambda)$ denotes the Fredho[lm] index of $T-\lambda$.

iv) If T is hyponormal, then $\frac{1}{2\pi i}\,dP$ is a positive measure.

Remarks: a) In addition to the above properties, it is implici[t] in the formula (1) that if we choose operators X',Y' in the algebra generated by X and Y, then there is a simple transformation by which we can obtain dP', the measure for X' and Y', from dP. This transformation law for dP plays a role in the proof of the theorem, and will be established in §7. It is also significant that dP does not transform like a measure, but more like a function.

b) After discovering the existence of dP, we surmised it must be essentially the same as the function $g(\mu,\nu)$ of Pincus [P_1], [P_2]. This has been varified by Pincus himself [P_3], who, making use of Wallach's lemma, has shown that formula (1') is valid for $dP = \frac{g(x,y)}{2\pi i}\,dxdy$. (Clearly formula (1') defines dP uniquely.) In terms of the $g(\mu,\nu)$ this paper establishes two properties of that function. One property is the transformation law referred to above. The other is the identification of $g(\mu,\nu)$ with the index of $T-(\mu+i\nu)$ off the essential spectrum of T. This had been done in certain cases by Pincus [P_2] and had been conjectured by him in general. Since learning of its general validity Pincus has developed interesting new proofs within the framework of his theory [$C\text{-}P_1$]. In addition to these properties, our result establishes the role $g(\mu,\nu)$ plays as an intrinsic feature of $\mathcal{P}\mathfrak{a}(X,Y)$. This was previously somewhat obscure.

c) The fact that dP is actually a measure and not a more singular distribution is of little importance to the main theme of this paper, and as the proof is technical we defer it to an appendix. On the other hand, various smoothness properties of dP are relevant in related contexts. We discuss these matters in §8.

d) We see that our bilinear form produces the integers associated by [B-D-F] to an operator normal modulo the compacts. In certain cases, $\langle\ ,\ \rangle$ is actually determined by these numbers. (see §7, prop. 7.2, corollary). However, in general it carries more information, and this extra information is wildly non-invariant under compact perturbation. We give examples in §9.

Also, we should mention that there is a 'determinant theory' parallel and actually equivalent to the trace theory we have presented here. This is discussed in §10.

The proof of the main theorem occupies paragraphs 2 through 6 below.

§1. An example: Toeplitz operators

Here we show by explicit computation that our main theorem holds for the case of Toeplitz operators. This result is implicit in Pincus $[P_2]$ but we believe the extreme elegance of the Toeplitz case has not been made explicit before. Putnam ($[Put_3]$, Chap.), gives an inequality for $tr[T^*,T]$, T a Toeplitz operator, in connection with adducing examples for his inequality $[Put_1]$, but he stops short of the full calculation. A somewhat more general computation than ours was made more or less simultaneously by Berger-Shaw [B-S].

We recall the definition of Toeplitz operators. Let $D \subseteq \mathbb{C}$ be the unit disk. Then ∂D, the boundary of D is $\mathbb{T}$, the unit circle. In $L^2(\mathbb{T})$ there is the space H^2 consisting of boundary values of functions holomorphic inside D. As is well-known, $L^2(\mathbb{T})$ has a natural orthonormal base $\{e_i\}_{i=-\infty}^{\infty}$ consisting of the characters of $\mathbb{T}$. With respect to this basis H^2 is the span of $\{e_i\}_{i=0}^{\infty}$. Let P be the orthogonal projection of L^2 onto H^2. Let f be a (continuous) function on $\mathbb{T}$. Then the Toeplitz operator with symbol f is the map $T_f : H^2 \to H^2$ defined by $T_f(g) = P(fg)$ for $g \in H^2$.

In terms of the basis $\{e_i\}_{i=0}^{\infty}$, the Toeplitz operator T_z is given by $T_z(e_i) = e_{i+1}$. In other words, T_z is just the shift The algebra of Toeplitz operators is thus the C^*-algebra generated by S. If $f = \sum_{i=-\infty}^{\infty} a_i e_i$ is the Fourier expansion of f, then $T_f = \sum_{i=0}^{\infty} a_i S^i + \sum_{i=1}^{\infty} a_{-i} S^{*i}$.

We see easily that $S^{*n}S^n = I$, while $S^nS^{*n} = I - P_n$ where P_n is the orthogonal projection onto the span of $\{e_i\}_{i=0}^{n-1}$. Thus $tr([S^{*n},S^n]) = n$. If $n \neq m$, then $tr([S^{*n},S^m]) = 0$. Now let $f = \sum_{i=-\infty}^{\infty} a_i e_i$ and $g = \sum_{j=-\infty}^{\infty} b_j e_j$ be two smooth functions on $\mathbb{T}$. Then we compute

$$[T_f,T_g] = \sum_{i,j\geq 0} (a_{-i}b_j - b_{-i}a_j)[S^{*i},S^j]$$. Taking traces we obtain

a) $$tr([T_f,T_g]) = \sum_{j\geq 0} j(a_{-j}b_j - a_jb_{-j}) = \sum_{j=-\infty}^{\infty} j\, a_{-j}b_j \ .$$

This formula may be interpreted in several ways. If θ is the standard variable on $\mathbb{T}$, so that $e_m = e^{im\theta}$, then we have

b) $$tr([T_f,T_g]) = \frac{1}{2\pi i}\int_{\mathbb{T}} f\, \frac{dg}{d\theta}\, d\theta \ .$$

Now let $\tilde{f}$ and $\tilde{g}$ be smooth extensions of f and g from $\mathbb{T} = \partial D$ to all of D . For example, $\tilde{f}$ and $\tilde{g}$ could be the harmonic functions in D with boundary values equal to f and g . By Stoke's theorem applied to the vector-valued function $(\tilde{f}\,\frac{\partial\tilde{g}}{\partial x}, f\,\frac{\partial\tilde{g}}{\partial y})$, we find that equation b) may be converted to

c) $$tr([T_f,T_g]) = \frac{1}{2\pi i}\int_D \left(\frac{\partial\tilde{f}}{\partial x}\frac{\partial\tilde{g}}{\partial y} - \frac{\partial\tilde{g}}{\partial x}\frac{\partial\tilde{f}}{\partial y}\right)dxdy$$

Formula c) amounts to our main theorem for the case of Toeplitz operators. For future reference, we note that the representing measure dP in this case is just $\frac{1}{2\pi i}\,dxdy$ on D (and zero off D.)

Finally, formula c) has a striking geometric interpretation. Supposing that $\tilde{f}$ and $\tilde{g}$ are real, we get a mapping $\phi : D \to \mathbb{R}^2$ given by $\phi(x,y) = (\tilde{f}(x,y),\tilde{g}(x,y))$; or by $\phi(z) = f(z) + ig(z)$. Then $\frac{\partial\tilde{f}}{\partial x}\frac{\partial\tilde{g}}{\partial y} - \frac{\partial\tilde{g}}{\partial x}\frac{\partial\tilde{f}}{\partial y}$ is just the Jacobian of this mapping. Thus, the integral in c) is just the area of $\phi(D)$. Here by area, we mean the area counted with multiplicities - a point $w \in \phi(D)$ is counted according to the number of times ϕ covers w in a positive sense minus the number of times ϕ covers D in a negative sense. Call this integer $\deg_\phi(w)$. Then we have

d) $\operatorname{tr}([T_f,T_g]) = \frac{1}{\pi}\int_{\phi(D)} \deg_\phi(w)\,dxdy$

Now it is known that $\deg_\phi(w)$ depends only on w and $\phi|_{\partial D}$, and is simply the winding number of the curve $\phi(\partial D)$ around w . On the other hand $\phi|_{\partial D}$ is just $f + ig$. From the theory of Toeplitz operators [D] we know that $\phi(\partial D)$ is just the essential spectrum of T_{f+ig} . Moreover, for $w \notin \phi(\partial D)$, the winding number of $\phi(\partial D)$ around w is just the Fredholm index of $T_{f+ig} - w$. Thus, for our final formula, we get the following beautiful relation between the index and self commutator of $T_{f+ig} - w$. Put $f + ig = h$.

e) $\operatorname{tr}([T_h^*,T_h]) = \frac{1}{2\pi i}\int_{\sigma(T_h)} \operatorname{ind}(T_h - w)\,dw\,d\bar{w}$.

In the next few sections, we will show these formulas hold much more generally.

§2. Wallach's Lemma

We now begin the proof of the main theorem by proving Wallach's lemma. Let (,) be a Pincus form on P_2 . We want to show (,) is of the form $(p,q) = \ell(J(p,q))$, for $p,q \in P_2$; hence forth we write $\ell \cdot J$ to mean $\ell(J(\ ,\ ))$. Here $J(p,q)$ denotes, as before, the Jacobian of p and q . First, let us note that $\ell \cdot J$ is indeed a Pincus form. If $p = p_1(r)$ and $q = q_1(r)$ for some $r \in S_2$, then we have

$$J(p,q) = \frac{\partial p}{\partial x}\frac{\partial q}{\partial y} - \frac{\partial q}{\partial x}\frac{\partial p}{\partial y} = \frac{dp_1}{dr}\frac{dq_1}{dr}\left(\frac{\partial r}{\partial x}\frac{\partial r}{\partial y} - \frac{\partial r}{\partial x}\frac{\partial r}{\partial y}\right) = 0 \quad .$$

Hence, certainly $\ell(J(p,q)) = 0$, and $\ell \cdot J$ is indeed collapsing. Next we observe that if $(,)_1$ and $(,)_2$ are two Pincus forms on P_2 , then $(\ ,\ )_3 = (,)_1 + (,)_2$ is also collapsing. Now define a linear functional ℓ on P_2 by the formula $\ell(\frac{\partial q}{\partial y}) = (x,q)$ for $q \in P_2$. We note that ℓ is indeed well defined since if $\frac{\partial q}{\partial y} = 0$, then q depends only on x, so $(x,q) = 0$ by the collapsing property.

Now consider the Pincus form $(,)_1 = (,) - \ell \cdot J$. By construction, $(,)_1$ has the property that $(x,q)_1 = 0$ for any $q \in P_2$. That is, x is in the radical of $(,)_1$. We claim this means that $(,)_1$ is identically zero. If so, then we have clearly proved the lemma. Our claim will follow in turn from two sublemmas.

Sublemma 1.: Let R be the radical of the Pincus form $(,)$. Then R is an algebra. In other words if $(p_i,q) = 0$ for all $q \in P_2$ and $i = 1,2$, then $p_1p_2 \in R$.

Proof: Take constants α and β, and any $q \in P_2$ and consider the equation

$$((\alpha p_1 + \beta p_2 + q), (\alpha p_1 + \beta p_2 + q)^2) = 0 ,$$

which follows from the collapsing property. Then, since $p_i \in R$, we see

$$(q, \alpha^2 p_1{}^2 + 2\alpha\beta p_1 p_2 + \beta^2 p_2^2 + 2\alpha p_1 q + 2\beta p_2 q + q^2) = 0 .$$

This equation is an identity in α and β. Taking the coefficient of $\alpha\beta$ we see $(q, p_1p_2) = 0$. Since q is arbitrary in P_2, we see $p_1p_2 \in R$, and the sublemma is proven.

Sublemma 2.: If $x \in R$, then $y \in R$.

Proof: If α is a constant, then by the collapsing property we have $(y+\alpha x, (y+\alpha x)^n) = 0$. Since $x \in R$, this gives $\sum_{i=0}^{n} \binom{n}{i} \alpha^i (y, x^i y^{n-i}) = 0$. Since this is an identity in α, we see $(y, x^i y^{n-i}) = 0$ for all $n \geq 0$ and $i \leq n$. Since the monomials $x^i y^{n-i}$ span P_2, we therefore have $y \in R$ as desired.

From these sublemmas, we see if $x \in R$, then R is a subalgebra containing x and y, or $R = P_2$, or $(,) \equiv 0$, and Wallach's lemma is proved.

§3. Functional calculus and extension of (,)

We return to our a.c. pair X,Y. In the introduction we defined the trace form $\langle , \rangle$ on $P\mathfrak{A}(X,Y)$. It would be nice to define it on $\mathfrak{A}(X,Y)$, but one obviously can't do this and has to compromise. We definitely need to extend the form $\langle , \rangle$ to a bigger subalgebra $\mathfrak{A}$ of $\mathfrak{A}(X,Y)$, and the form $(,)$ to a bigger function algebra on the plane. Surprisingly enough this works out rather well: the appropriate algebras are closed under many operations and the bilinear form $(,)$ extends uniquely to a continuous form on $C^{\infty}(\mathbb{R}^2)$ with compact support (and fairly low order). Actually we work with $S(\mathbb{R}^2)$ rather than $C^{\infty}(\mathbb{R}^2)$ and view $(,)$ as a distributional bilinear form. There are several reasonable ways to present the material in this section. One would be to present the minimal functional calculus needed in the rest of the paper, thereby making the proofs simple. The other is to set down a very general and flexible functional calculus which one can use for most circumstances without much understanding of the proofs. We have opted for the second course. Thus, this section is quite technical, but the casual reader may skip all but the statements of definitions and the main theorems 3.1 and 3.3 and 3.4.

There are several reasonable ways to construct a functional calculus for several operators: one is to take limits of polynomials in these operators, another is to take limits of trigonometric polynomials. We will need both methods. Actually our constructions will be valid for a class of operators more general than self-adjoint operators. Although we make no use of this extra generality, we mention it to indicate that many of our considerations extend naturally to a Banach space context.

Let $\mathfrak{A}(X,Y) = \mathfrak{A}$ be the C*-algebra generated by X and Y. Choose $\mathfrak{A}_1 \subseteq \mathfrak{A}$ as follows. It is a self-adjoint set in $\mathfrak{A}$, contains

X and Y, and such that for any two elements $A,B \in \mathfrak{A}_1$, the commutator $[A,B]$ is trace class. Secondly, it is a maximal set with these properties. Such an $\mathfrak{A}_1$ exists by the obvious Zorn's lemma argument. We will show $\mathfrak{A}_1$ is in fact a *-algebra and closed with respect to certain operations, which we now detail.

Let $\{A_i\}_{i=1}^n$ be operators. Suppose each A_i has operator norm $||A_i|| < r$ for some fixed r. Consider power series $\Sigma a_\alpha x_1^{\alpha_1} \ldots x_n^{\alpha_n}$ in variables $\{x_i\}_{i=1}^n$. Here $\alpha = (\alpha_1, \alpha_2, \ldots, \alpha_n)$ is a "multi-index" - an n-tuple of non-negative integers. We put $|\alpha| = |\alpha_1| + |\alpha_2| + \ldots + |\alpha_n|$. Define a norm $|| \; ||_r$ on these series by $||\Sigma a_\alpha x_1^{\alpha_1}, \ldots, x_n^{\alpha_n}||_r = \Sigma |a_\alpha| r^{|\alpha|}$. Let $PS(r)$ denote the space of series for which $|| \; ||_r$ is finite and consider it as a Banach space with this norm. Then by our assumption on the A_i, the mapping e_A defined by $e_A(\Sigma a_\alpha x_1^{\alpha_1} \ldots x_n^{\alpha_n}) = \Sigma a_\alpha A_1^{\alpha_1} \ldots A_n^{\alpha_n}$ is continuous from $PS(r)$ to $L(H)$. We call the range of e_A the <u>Wick-ordered power series operational calculus</u> of radius r in $A_1, \ldots, A_n$.

Now suppose the A_i are self-adjoint. Consider

$\exp_A(s_1, \ldots s_n) = e^{is_1A_1} e^{is_2A_2}, \ldots, e^{is_nA_n}$. Then $\exp_A$ is a holomorphic function from $\mathbb{C}^n$ to $L(H)$ and is unitary valued for real $s = (s_1, \ldots, s_n)$. Let $S(\mathbb{R}^n)$ be the Schwartz space of $\mathbb{R}^n$ - the space of all rapidly decreasing smooth functions. We define a map

$$\wedge_A : S(\mathbb{R}^n) \to L(H) \text{ by } \wedge_A(f) = \int_{\mathbb{R}^n} f(s) \exp_A(s) ds \; .$$

Then it is standard [C-F] that $\wedge_A$ defines a continuous map from $S(\mathbb{R}^n)$. We call the range of $\wedge_A$ the <u>Wick-ordered</u> C^∞ <u>functional calculus</u> in $A_1, \ldots, A_n$. We will also write $\wedge_A(f) = \hat{f}(A_1, \ldots, A_n)$.

Remark: Let $f \in S(\mathbb{R}^n)$ be a product, $f(s_1, \ldots, s_n) = f_1(s_1) \ldots f_n(s_n)$. Then $\wedge_A(f)$ is a product also: $\wedge_A(f) = \prod_{i=1}^n \wedge_{A_i}(f_i)$. Now if the ordinary Fourier transformation $\hat{f}_i$ of f_i has support disjoint from $\sigma(A_i)$, then it is obvious that $\wedge_{A_i}(f_i) = 0$. Thus $\wedge_A(f) = 0$. Therefore, if $||A_i|| < r$ for each i, and any f_i is such that $\hat{f}_i$ has support outside $(-r,r)$, then $\wedge_A(f) = 0$. By continuity, it follows that if $\hat{f}$ has support outside the box $B_r = \{(s_1,\ldots,s_n); |s_i| < r\}$, then $\wedge_A(f) = 0$. In other words, $\wedge_A(f)$ depends only on the restriction of $\hat{f}$ to B_r. Since any analytic function on B_r may be extended to a Schwartz function on $\mathbb{R}^n$, it follows that the Wick-ordered C^∞ functional calculus contains the Wick-ordered power series functional calculus for $A_1,\ldots,A_n$, although formally it does not. Also, we may formally extend $\wedge_A$ from $S(\mathbb{R}^n)$ to all distributions which are Fourier transforms of smooth functions, with the understanding that $\wedge_A(S)$ for such a distribution S is just $\wedge_A(f)$ for any $f \in S$ such that $\hat{S}-\hat{f}$ vanishes on B_r. This linguistic convention will be convenient at times.

In these terms we may state our main result on the functional calculus.

Theorem 3.1: Let X,Y be an a.c. pair. Let $\mathfrak{A}_1 \subseteq \mathfrak{A}(X,Y)$ be a maximal self-adjoint set containing X and Y and such that any pair of elements of $\mathfrak{A}_1$ have trace class commutator. Then

i) $\mathfrak{A}_1$ is an algebra. In particular $\mathfrak{A}_1 \supseteq P\mathfrak{A}(X,Y)$.

ii) If $\{A_i\}_{i=1}^n \subseteq \mathfrak{A}_1$, and $||A_i|| < r$ for all i, then $\mathfrak{A}_1$ contains the Wick-ordered power series operational calculus of radius r in the A_i.

iii) If $\{A_i\}_{i=1}^n \subseteq \mathfrak{A}_1$ are self-adjoint, then $\mathfrak{A}_1$ contains the C^∞ Wick-ordered functional calculus in the A_1. In particular, if $B \in \mathfrak{A}_1$ is Fredholm, then $\mathfrak{A}_1$ contains both factors in the polar decomposition of B.

The proof requires some preparation. We begin by recalling the basic facts on trace class operators. Let T be the set of trace class operators in $L(H)$. Then T is a *-ideal in $L(H)$, and has a natural norm $|| \; ||_1$, the trace norm, with respect to which T is complete. For $A \varepsilon L(H)$, $B \varepsilon T$, the inequalities $||AB||_1 \leq ||A|| \cdot ||B||_1 \geq ||BA||_1$ hold. Any trace class operator has a trace. If $\{e_i\}_{i=1}^{\infty} \subseteq H$ is any orthonormal basis, and $B \varepsilon T$, and $\{b_{ij}\}$ is the matrix of B with respect to the $\{e_i\}$, then we may compute $\operatorname{tr} B$ by the formula $\operatorname{tr} B = \sum_{i=1}^{\infty} b_{ii}$. If B is positive then $\operatorname{tr} B = ||B||_1$.

The basic identity on which theorem 3.1 is based is

$$[A,BC] = B[A,C] + [A,B]C \quad , \tag{3.1}$$

expressing the fact that $B \to [A,B]$ is a derivation. This immediately shows $\mathfrak{A}_1$ is an algebra. Indeed, if $A,B,C \varepsilon \mathfrak{A}_1$, then $[A,BC] \varepsilon T$ since T is an ideal. Similarly, $[A,(BC)^*] = [A,C^*B^*] \varepsilon T$, since $\mathfrak{A}_1$ is self-adjoint. Finally, $[BC,(BC)^*]$ is trace class by a repetition of (3.1). Thus $\mathfrak{A}_1 \cup \{BC,(BC)^*\}$ again satisfies the conditions on $\mathfrak{A}_1$, and so $BC \varepsilon \mathfrak{A}_1$ by maximality. Since $\mathfrak{A}_1$ is even more easily seen to be a linear space, it is an algebra. A more quantitative analysis along the same lines will give us ii) and iii) of theorem 3.1.

Remark: Let $\{A_i\}_{i=1}^{n} \subseteq \mathfrak{A}_1$ be a finite set. Consider the quantity $||B||' = ||B|| + \max_i ||[B,A_i]||_1$ for $B \varepsilon \mathfrak{A}_1$. The identity (3.1) lets one show $||B||'$ is an algebra norm on $\mathfrak{A}_1$, i.e., $||B_1B_2|| \leq ||B_1||' ||B_2||'$. There are many analogies between $|| \; ||'$ and the C^1 norm on functions (on say an interval $[a,b] \subseteq \mathbb{R}$). It might be interesting to consider the Banach algebras obtained from $\mathfrak{A}_1$ by completing with respect to $|| \; ||'$, or the locally m-convex algebra structure defined on $\mathfrak{A}_1$ by all the norms $|| \; ||'$.

A reiteration of (3.1) yields

$$[A, \prod_{i=1}^{n} B_i] = \sum_{j=1}^{n} \prod_{i<j} B_i [A,B_j] \prod_{j<i} B_i \quad . \tag{3.2}$$

If $||B_i|| \leq r$ for all i, then (3.2) in turn yields the estimate

$$(3.3) \qquad ||[A, \prod_{i=1}^{n} B_i]||_1 \leq n \max_j ||[A,B_j]|| \, r^{n-1} .$$

If A is also a product, say $A = \prod_{i=1}^{m} A_i$, and $||A_i|| \leq r$, then repeating (3.3) gives

$$(3.4) \qquad ||[\prod_{i=1}^{m} A_i, \prod_{j=1}^{n} B_j]||_1 \leq nm \max_{i,j} ||[A_i,B_j]|| r^{n+m-2} .$$

Now choose $A_1,A_2,\ldots,A_n \subseteq \mathfrak{A}_1$, and take $r > \max||A_i|| = \rho$. Consider a power series $s = \Sigma\, a_\alpha x_1^{\alpha_1} \ldots x_n^{\alpha_n}$ in $PS(r)$. We want to show $e_A(s) \in \mathfrak{A}_1$. To do so, it is sufficient to show that $[e_A(s),B] \in T$ and $[e_A(s),e_A(s)^*] \in T$. By (3.3) we have $||[e_A(s),B]||_1 \leq (\Sigma_\alpha |a_\alpha||\alpha|\rho^{|\alpha|-1}) \max_i ||[A_i,B]||_1$. Since $\rho < r$, the series in parentheses is absolutely convergent, so the partial sums of $[e_A(s),B]$ converge absolutely in T, so $[e_A(s),B]$ is indeed trace class. The other commutator is handled similarly. We have

$$||[e_A(s),e_A(s)^*]|| \leq (\sum_{\alpha,\beta} |a_\alpha||a_\beta||\alpha||\beta|\rho^{|\alpha|+\beta-2}) \max_{i,j} ||[A_i,A_j^*]||_1 .$$

The series in parentheses is the square of the derivative of $\Sigma|a_\alpha|x^{|\alpha|}$ at $x = \rho$, and so is absolutely convergent. Thus $[e_A(s),e_A(s)^*] \in T$, and we have demonstrated point ii) of Theorem 3.1.

We have also almost proven the following lemma.

<u>Lemma 3.2.</u> Take $A_1,A_2,\ldots,A_n,B_1,\ldots,B_m, C \in \mathfrak{A}_1$. The mappings

$$\phi : (s_1,\ldots,s_n) \to [\prod_{j=1}^{n} e^{is_jA_j}, C] \quad \text{and}$$

$$\psi : (s_1,\ldots,s_n, t_1,\ldots,t_n) \to [\prod_{j=1}^{n} e^{is_jA_j}, \prod_{k=1}^{m} e^{it_kB_k}]$$

are holomorphic mappings from $\mathbb{C}^n$ and $\mathbb{C}^{n+m}$ respectively into T. Moreover, if $A_1,\ldots,A_n,B_1,\ldots,B_m$ are Hermitian, then $||\phi(s)||_1 \leq \beta(|s|+1)$ and $||\psi(s,t)||_1 \leq \beta(|s|+1)(|t|+1)$ for some constant β and $s \in \mathbb{R}^n$, $t \in \mathbb{R}^m$. Here $|s| = \sum_{i=1}^{n} |s_i|$.

Proof: The holomorphicity is quite clear from the estimates performed above. Let us prove the bounds in the Hermitian case. We have $||\phi(s)||_1 \leq n \max_j ||[e^{is_jA_j}, C]||_1$ and $||\psi(s,t)||_1 \leq$

$nm \max_{j,k} ||[e^{is_jA_j}, e^{it_kB_k}]||_1$ by (3,3) and (3.4). Thus it suffices to consider the case when $n = m = 1$. Thus, set $A_j = A$, $s_j = s$, $B_k = B$ and $t_k = t$. Since $[-1,1]$ is compact, there is some constant β_0 such that $||[e^{isA}, C]||_1 \leq \beta_0$ for $|s| \leq 1$. Then, if $n - 1 < |s| \leq n$, we have again by (3.3)

$$||[e^{isA}, C]||_1 = ||[(e^{i\frac{s}{n}A})^n, C]||_1 \leq n||[e^{i\frac{s}{n}A}, C]||_1 \leq n\,\beta_0 \leq \beta(|s|+1)\ .$$

Likewise, another application of (3.4) does the trick for ψ.

Now we turn to point iii) of theorem 3.1. Consider self-adjoint elements $\{A_i\}_{i=1}^n \subseteq \mathfrak{A}_1$, and $f \in S(\mathbb{R}^n)$; we will show we could add $\wedge_A(f)$ to $\mathfrak{A}_1$ and preserve the relevant properties. This will imply $f \in \mathfrak{A}_1$ already. If $C \in \mathfrak{A}_1$ then

$$[\wedge_A(f), C] = \int_{\mathbb{R}^n} f(s)[\prod_{i=1}^{n} e^{is_iA_i}, C] = \int_{\mathbb{R}^n} f(s)\phi(s)ds,$$ where ϕ is as in Lemma 3.2. The estimate in that lemma shows $f(s)\phi(s)$ is an absolutely integrable function from $\mathbb{R}^n$ to T. Thus, $[\wedge_A(f), C]$ is trace class, as desired. Similarly, the estimate on the function ψ in lemma 3.2, with $B_i = A^*_{n-i}$, shows $[\wedge_A(f), \wedge_A(f)^*] \in T$. This proves the first statement in iii).

To finish the theorem, consider some Fredholm $A \in \mathfrak{A}_1$. Let $A = UP$ be the polar decomposition of A. Then P is a positive Fredholm operator, $P = (A^*A)^{1/2}$, and U is a Fredholm partial isometry such that $\ker U^* = \ker P$. Since A^*A is positive and Fredholm, we may find B in the C^∞ functional calculus of A^*A such that $PB = Q$ is projection onto the range of P. Then we have $AB = UPB = UQ = U \in \mathfrak{A}_1$. This concludes theorem 3.1.

We now give some more details on the C^∞ functional calculus. Again take self-adjoint elements $\{A_i\}_{i=1}^n \subseteq \mathfrak{A}_1$. Let $\wedge : S(\mathbb{R}^n) \to S(\mathbb{R}^n)$ be the usual Fourier transform (appropriately normalized). We define $\tau_A : S(\mathbb{R}^n) \to \mathfrak{A}_1$ by $\tau_A(f) = \wedge_A(\hat{f})$.

Proposition 3.3: The map τ_A has the following properties:

i) $\tau_A(f)^* - \tau_A(\bar{f}) \in T$. Here $\bar{f}$ is the complex conjugate of f . Moreover, the map $f \to \tau_A(f)^* - \tau_A(\bar{f})$ is continuous (and complex antilinear) from S to T .

ii) $\tau_A(f)\tau_A(g) - \tau_A(fg) \in T$, and the corresponding map is bicontinuous in f and g .

iii) $[\tau_A(f), \tau_A(g)] \in T$, and again is bicontinuous in f and g .

Proof: All these statements follow from Lemma 3.2, which gives the behaviour of the Fourier transforms of the kernels of these maps. We see iii) is immediate from ii). Thus we will do ii) and leave i) to the reader.

We know $\widehat{fg} = \hat{f} * \hat{g}$, the $*$ denoting convolution. Thus, we must show $\wedge_A(f)\wedge_A(g) - \wedge_A(f*g) \in T$. But

$\wedge_A(f)\wedge_A(g) = (\int_{\mathbb{R}^n} f(s)\exp_A(s)ds)(\int_{\mathbb{R}^n} g(t)\exp_A(t)dt) =$

$\int_{\mathbb{R}^{2n}} f(s)g(t)\exp_A(s)\exp_A(t)dsdt$, by Fubini. On the other hand

$\wedge_A(f*g) = \int_{\mathbb{R}^n}(f*g)(s)\exp_A(s)ds = \int_{\mathbb{R}^n}(\int_{\mathbb{R}^n} f(s-t)g(t)dt)\exp_A(s)ds =$

$\int_{\mathbb{R}^{2n}} f(s)g(t)\exp_A(s+t)dsdt$, again by Fubini. Thus, the difference is given by

$$\int_{\mathbb{R}^{2n}} f(s)g(t)(\exp_A(s)\exp_A(t) - \exp_A(s+t))dsdt \ .$$

But we have $\exp_A(s)\exp_A(t) - \exp_A(s+t) =$

$(\prod_{j=1}^n e^{is_jA_j})(\prod_{k=1}^n e^{it_kA_k}) - (\prod_{j=1}^n e^{i(s_j+t_j)A_j})$. If $e^{is_jA_j} = U_j$ and $e^{it_jA_j} = V_j$, then the above difference is

$$(\pi U_j)(\pi V_k) - \pi(U_j V_j) = \sum_j (\pi_{k<j} U_k)[U_j, \pi_{k<j} V_k](\pi_{j<k} U_k V_k) \quad .$$

By Lemma 3.2, this is an analytic, slowly growing kernel on $\mathbb{R}^{2n}$, so integration against it does give a good T-valued bilinear form on S.

Now consider on $\mathfrak{A}_1$ the bilinear form $<,>$ defined by $<A,B> = tr([A,B])$, for $A,B \in \mathfrak{A}_1$. We will call this the canonical bilinear form on $\mathfrak{A}_1$. If $\{A_i\}_{i=1}^n \subseteq \mathfrak{A}_1$ are Hermitian, then we have the following corollary of Proposition 3.3, iii):

<u>Corollary</u>: The bilinear form $<,> \cdot \tau_A$ is a continuous bilinear form on $S(\mathbb{R}^n)$. It has compact support, and so has a formal extension to $C^\infty(\mathbb{R}^n)$.

The statement about compact support follows because τ_A itself has compact support, as we mentioned when introducing the C^∞ functional calculus.

Now we specialize to the case $A_1 = X$, and $A_2 = Y$, and we refer to the map τ_A simply as τ. Then we have the formal extension $\tau : C^\infty(\mathbb{R}^2) \to \mathfrak{A}_1$, and as we have seen, the image of τ contains the subspace of $P\mathfrak{A}(X,Y)$ consisting of Wick-ordered elements. Clearly the pullback of $<,>$ on these elements gives rise to the bilinear form $(,)$ on P_2 defined in the introduction. As we have said, this form is collapsing (see also Proposition 4.1). Therefore, so also will be $<,> \cdot \tau$, its continuous extension to C^∞. Now by Wallach's lemma $(,)$ is represented in the form $\ell \cdot J$ where $J(f,g) = \frac{\partial f}{\partial x}\frac{\partial g}{\partial y} - \frac{\partial g}{\partial x}\frac{\partial f}{\partial y}$ for $f,g \in P_2$. If we can show ℓ has a continuous extension to C^∞, then $<,> \cdot \tau$ will also have such a representation by continuity. Recall that ℓ was defined by $\ell(\frac{\partial f}{\partial y}) = (x,f)$. Using this definition, we can extend ℓ to C^∞. Then since $f \to \frac{\partial f}{\partial y}$ is a surjective open mapping from C^∞ to itself, we see this extended ℓ will be a distribution (of compact support). We summarize this discussion in the following theorem.

<u>Theorem 3.4</u>: The bilinear form $(,)$ defined on P_2 in the

introduction, has an unique continuous extension, still denoted (,) to $C^\infty(\mathbb{R}^2)$. This extended form has the collapsing property and has the representation

$$(f,g) = \ell(J(f,g))$$

where $J(f,g)$ is the Jacobian of f and g and ℓ is a distribution of compact support. For $f,g \in S$, we also have

$$(f,g) = \int_{\mathbb{R}^2\times\mathbb{R}^2} \hat{f}(s,t)\hat{g}(u,v)\,\mathrm{tr}([e^{isX}e^{itY},e^{iuX}e^{ivY}])\,ds\,dt\,du\,dv$$

where $\hat{f},\hat{g}$ are the Fourier transforms of f and g .

We will call ℓ the representing functional for (,) .

Remarks: a) We have not attempted to be precise about the exact degree of continuity of τ_A and $\langle\ ,\ \rangle\cdot\tau_A$. Lemma 3.2 however, shows it is very low-roughly separately C^1 , except for the usual ambiguity depending on the number of variables. Thus, it is plausible ℓ could be a measure.

b) The C^∞ functional calculus (at least for everything in it not involving adjoints) clearly could have been performed for operators $\{A_i\}_{i=1}^n$ satisfying $||e^{isA_i}|| \leq \beta(|s|+1)^\lambda$ for some constants $\beta \geq 1$, $\lambda \geq 0$. These are so-called operators with $\mathbb{C}^n$-functional calculus and exist in a general Banach space context (see [C-F]).

§4. The trace form and the essential spectrum

In the preceding section we showed that $\langle f,g\rangle$ could actually be defined for $f,g \in S(\mathbb{R}^2)$. In this paragraph we show that $\langle f,g\rangle$ is actually determined by the restriction of f and g to the joint essential spectrum E of X,Y . Recall that $E = \{(x,y): x+iy \in e\sigma(T)\}$. Actually, we use E and $e\sigma(T)$ interchangeably throughout the paper. As in §3 we denote the functional calculus mapping $\tau_A(f)$ of Proposition 3.3 by $\tau(f)$ when $A = \{X,Y\}$. It is obvious from the Gelfand construction that the operator $\tau(f)$ in $\mathfrak{A}(X,Y)$ becomes $f|_E$ under the Gelfand map. The following is a basic but simple fact.

Proposition 4.1: Suppose A and B in $L(H)$ have trace class commutator. Suppose A is normal and diagonalizable, i.e., A has a complete orthonormal system of eigenvectors. Then $tr([A,B]) = 0$. In particular, if A is compact and normal , then $tr([A,B]) = 0$.

Proof: Let $C = [A,B]$ and let $\{c_{ij}\}$ be the matrix of C with respect to the orthonormal basis for A . Then $trC = \Sigma\, c_{ii}$; but in this basis $c_{ii} \equiv 0$ for all i , just as it does in the finite dimensional case. One can easily compute this.

Remark: This result is valid in more general contexts. The raison d'etre for this section is

Corollary 1. If f,g,f_1,g_1 are in $S(\mathbb{R}^2)$ and if $f = f_1$ on E and $g = g_1$ on E , then $(f,g) = (f_1,g_1)$. In particular, the form $(\, , \,)$ is supported on E .

Proof: Set $h = f - f_1$. Since h vanishes on E , the operator $\tau(h)$ has Gelfand representation identically equal to zero. Thus it is compact. Since $\tau(h)^*$ is in $\mathfrak{A}_1$ the operators $\tau(h)^* \pm \tau(h)$ and $\tau(g)$ have trace class commutator. Proposition 4.1. implies $(f,g) = (f_1,g)$. The same argument for $g-g_1$ proves the corollary.

The importance of this is that though our entire set up depends on trace class commutors and a C^∞-functional calculus, the basic form $< \, , \, >$ depends only on the 'symbol' of the operators involved. For the sake of completeness we insert a more algebraic statement of the corollary.

Corollary 2. Let $\mathfrak{A}_1$ be a (not necessarily closed)*-algebra of operators whose commutator ideal consists of trace class operators. Consider the canonical bilinear form $<A,B> = tr([A,B])$ on $\mathfrak{A}_1$. Then $\mathfrak{A}_1 \cap LC(H)$ is in the radical of $<,>$. Hence $< \, , \, >$ factors to a form on the abelian algebra $\mathfrak{A}_1/\mathfrak{A}_1 \cap LC(H)$.

§5. The Fredholm index

Corollary 1 of the preceding section has a striking consequence: the representing functional ℓ is locally constant on the complement of the joint essential spectrum E of X and Y. That is

Proposition 5.1. If U is a component of the complement of E, then there is a constant K so that

$$(5.1) \qquad \langle f,g\rangle = K \int J(f,g)\,dxdy$$

whenever f or g has support in U.

Proof of Proposition: If $f \in S(\mathbb{R}^2)$ is supported in U, then it vanishes on E and so

$$\ell\left(\frac{\partial f}{\partial x}\right) = \langle f,y\rangle = 0$$

that is the distributional derivative of ℓ on U in the y direction vanishes identically. Likewise for $\frac{\partial}{\partial x}$. Thus, in the sense of distributions ℓ on U is a constant. In this section we evaluate the constant K.

Proposition 5.2. The constant K is $-1/2\pi i$ times the Fredholm index of $(X+iY - (x_0+iy_0)I)$ for any point (x_0,y_0) in U.

Proof: Since translating X and Y by $x_0 I$ and $y_0 I$ does not affect this problem we may assume that U contains the origin. Write the polar decomposition of $T = X+iY$ as RV where R is non-negati[ve] and V is a partial isometry. Since T is Fredholm, Theorem 3.1 implies that R and V belong to our functional calculus $\mathfrak{A}_1$ and that if γ is any C^∞-function $\gamma(R)$ belongs to the functional calculus. Thus we have

$$(5.1) \qquad \mathrm{tr}[(\gamma(R)V)^*,\ \gamma(R)V] = \ell(J(\gamma(r)e^{-i\theta},\gamma(r)e^{i\theta}))$$

Since T is Fredholm, R is Fredholm, so there is an interval $(0,\delta]$ which does not intersect $\sigma(R)$. If γ is identically equal to 1 on $(\delta, ||R||]$, then $\gamma(R)$ is equal to the projection onto range V : consequently, $\gamma(R)V = V$. For such a γ the left hand side of the equation above equals index V = index T . If δ is small enough so that the circle of radius δ is inside of U , then the right side of the equation is $K \int J(\gamma(r)e^{-i\theta}, \gamma(r)e^{i\theta})dxdy$. To evaluate this choose $\gamma(r) = 1/\delta\ r$ on $[0,\delta]$; one gets $K/\delta^2 \int_\Delta J(x-iy, x+iy)dxdy$ where Δ is the disk of radius δ . This is equal to $2\pi iK$. One might observe that the function γ we have used is not smooth and is therefore not admissible. However, this poses no problem since γ can be approximated by smooth functions. End of proof.

We conclude this discussion by remarking that once one knows the representing functional ℓ extends to $S(\mathbb{R}^2)$, then a direct proof of the index theorem can be gotten quickly by performing one simple estimate of section 3 type. This is because it suffices to prove that if (x_0, x_0) is any point of U and if χ is the characteristic function of any small disk Δ around (x_0, y_0) , then $\ell(\chi) = 2\pi i$ index $[X+iY - (x_0+iy_0)I]$. We may choose $(x_0, y_0) = (0,0)$ as before. All that we have to do is perform a simple estimate which gives (5.1) directly.

This index result has a striking consequence. If A belongs to $\mathfrak{A}(X,Y)$, then the Gelfand map applied to $\mathfrak{A}(X,Y)/LC\cap\mathfrak{A}(X,Y)$ associates with A a function on $C(E)$ called the <u>symbol</u> of A .

<u>Proposition 5.3</u>. Suppose that X,Y are an almost commuting pair. Then the index of an operator V in $\mathfrak{A}(X,Y)$ with modulus one symbol $\tilde{v}$ having an extension to a function v in $C^4(\mathbb{R}^2)$ is

$$\text{index } V = \frac{1}{2\pi i} \text{ index } (T-\lambda_i I) \int_{U_j} J(v,\tilde{v})dxdy$$

where the U_j are components of E complement and the λ_j are points in U_j .

<u>Proof</u>: Since the symbol $\tilde{v}$ is modulus 1 on E and has an extension

to a function in $C^4(\mathbb{R}^2)$ a straight forward argument shows that it has an extension to a function w in $C^4(\mathbb{R}^2)$ which has modulus = 1 on and near E. The operators V, $\tau(v)$, and $\tau(w)$ differ by a compact operator and consequently they have the same index. By the representation theorem

$$\text{index } V = \text{tr}[\tau(w),\tau(w)^*] = \frac{1}{2\pi i}\,\text{index}(T-\lambda_j I)\int_{U_j} J(w,\bar{w})\,dxdy$$

and since w and v have the same values on the boundary of U_j we get the proposition.

§6. Putting together the pieces

In this section we prove Theorem 1. First we must define the planar measure P. If $[a,b)$ and $[c,d)$ are half open intervals, define $E_{[a,b)}$ to be the spectral projection for X associated with the interval $[a,b)$, and $F_{[c,d)}$ to be the spectral projection for $E_{[a,b)}YE_{[a,b)}$ associated with $[c,d)$. Note: $F_{[c,d)} = E_{[a,b)}F_{[c,d)}E_{[a,b)}$. We define a set function $\tilde{P}$ on the semi-algebra $\mathcal{K}$ of half open and closed rectangles $[\ ,\ ) \otimes [\ ,\ )$ by

$$\tilde{P}([a,b) \otimes [c,d)) = \text{tr}(E_{[a,b)}F_{[c,d)}E_{[a,b)}[X,Y])$$

One can prove

<u>Proposition 1.6.1</u>. The set function $\tilde{P}$ is finitely additive and extends to a regular signed measure P of bounded total variation such that

$$\ell(f) = \int f dP$$

Moreover, if T is hyponormal then P is positive.

<u>Proof</u>: See Appendix to Section 1.

We are now ready to discuss Theorem 1. Parts (ii) and (iii) of Theorem 1 are obvious consequences of this proposition and subsection 4 and 5 respectively. Part (i) is an obvious consequence of part (iii). Part (iv) is contained in the last line of Proposition 1.6.1 and is obvious from the definition of $\tilde{P}$. Theorem 1 is finished.

§7. The transformation law

Let f and g be real-valued C^∞ functions on $\mathbb{R}^2$. Then we can form the operators $\tau(f)$ and $\tau(g) \in \mathfrak{A}_1$ using the Wick-ordered C^∞ functional calculus for X and Y . According to proposition 3.3, $\tau(f)^* - \tau(f)$ and $\tau(g)^* - \tau(g)$ are trace class, since f and g are real. Put $R = (\frac{1}{2})(\tau(f)+\tau(f)^*)$ and $S = (\frac{1}{2})(\tau(g)+\tau(g)^*)$. Let τ' be the Wick-ordered C^∞ functional calculus in R and S . Then $\tau' : C^\infty(\mathbb{R}^2) \to \mathfrak{A}_1$, and if $(\, , \,)' = \langle \, , \, \rangle . \tau'$ where $\langle \, , \, \rangle$ is the canonical form on $\mathfrak{A}_1$, then $(\, , \,)'$ is a Pincus form, and has a representative functional ℓ' . We want to relate ℓ' to the representative functional ℓ of the form $(\, , \,)$ on $C^\infty(\mathbb{R}^2)$ attached to X and Y by theorem 3.4.

We can interpret f and g as defining a map on $\mathbb{R}^2$. Thus define $\theta : \mathbb{R}^2 \to \mathbb{R}^2$ by $\theta(x,y) = (f(x,y),g(x,y))$. Then we have the dual map $\theta^* : C^\infty(\mathbb{R}^2) \to C^\infty(\mathbb{R}^2)$, given by $\theta^*(h) = h \cdot \theta$

<u>Proposition 7.1</u>: The representing functionals ℓ' and ℓ are related by the formula

(7.1) $\qquad \ell'(h) = \ell(\theta^*(h)\Delta(\theta))$, where $\Delta(\theta) = J(f,g)$ is the is the Jacobian of θ .

<u>Proof</u>: Let h_1 and h_2 be Wick-ordered polynomials in R and S . Then we have $\ell'(J(h_1,h_2)) = \langle J'(h_1),J'(h_2)\rangle = \mathrm{tr}([h_1(R,S),h_2(R,S)])$. On the other hand, by Proposition 3.3, we see that $h_i(R,S)-h_i\cdot\theta(X,Y)$ is trace class. Therefore, by Proposition 4.1, we have $\langle\tau'(h_1),\tau'(h_2)\rangle = \langle\tau(h_1\cdot\theta),\tau(h_2\cdot\theta)\rangle = \ell(J(h_1\cdot\theta,h_2\cdot\theta))$. By the chain rule for Jacobians $J(h_1\cdot\theta,h_2\cdot\theta) = (J(h_1,h_2)\cdot\theta)J(f,g)$. Set $J(h_1,h_2) = h$, and observe that we have proven (7.1) for all polynomials h . By continuity, then, it holds for all $h \in C^\infty$,

and the proposition is proved.

Remarks. a) Although we have nominally proved (7.1) for all $f,g \in C^\infty$ one sees that ℓ' actually depends on θ only on support of ℓ, a compact set. Thus, if we had restricted to $f,g \in S(\mathbb{R}^2)$ there would have been no loss of generality: the extension to all C^∞ is purely formal.

b) Along the same lines, note that the transformation law is local in the sense that the behavior of ℓ' on U (U open $\subseteq \mathbb{R}^2$) depends only on ℓ restricted to $\theta^{-1}(U)$.

c) As we remarked in the introduction, ℓ does not transform as a measure. The transformation law for a measure would not require the multiplication of $\theta^*(h)$ by $\Delta(\theta)$. This will be taken up again in Part II.

To give a better idea of the law, consider the case when ℓ is just integration against Lebesgue measure over some open set U with smooth boundary ∂U. This case is particularly relevant to us in view of part ii) of our main theorem and the locality property mentioned in Remark b) above. In this case ℓ' is given by $\ell'(f) = \int_U (f\cdot\theta)\Delta(\theta)dxdy$. If $\theta : U \to \theta(U)$ is a smooth orientation preserving diffeomorphism, then the change of variables law for multiple integrals says ℓ' is just integration over $\theta(U)$, while if θ is orientation reversing, then it is minus the integral over $\theta(U)$. In general, we see each point v not in $\theta(\partial U)$ will be covered a certain number of times in a positive sense by θ, and a certain number of times in a negative sense. Let $\deg_\theta(v)$ denote the algebraic sum of the times v is covered by θ. Then by breaking up U into small sets on each of which θ is a diffeomorphism, using the above conclusion and summing by an appeal to locality, we conclude

$$\ell'(f) = \int_{\theta(U)} f(v)\deg_\theta(v)dv_1dv_2 .$$

We now give an alternate proof of the index result using the transformation law. The proof proceeds by reducing to the case when $T = X + iY$ is unitary modulo the compacts, and then comparing with the shift, for which we have computed everything explicitly in §1. There is of course considerable overlap with the proof already given. This proof is somewhat longer, but has the virtue of taking maximum advantage of the considerable formal structure available in the situation and thus it has a better chance of extending to more difficult problems.

As in the previous proof of the index theorem, it suffices to assume that $0 \notin E$ and to evaluate the constant given by our representation near 0.

Let B_δ be the disk of radius δ about 0. Choose $0 < \delta < 1$ so that $B_\delta \cap E$ is empty. Let f be a positive C^∞ function on $\mathbb{R}$ such that $f(r) \equiv 1$ for $r \leq \delta/2$, $f(r) = 1/r$ for $r \geq \delta$ and $r\,f(r)$ is monotone increasing. Define g on $\mathbb{C}$ by $g(z) = f(|z|)z$. Then g is a diffeomorphism near 0 and has absolute value identically one on E.

Consider $\tau(g)$. Then $e\sigma(\tau(g)) = g(e\sigma(T))$, so g is unitary modulo the compacts. Moreover, if $g_t(z) = z(1-t + tf(|z|))$, then $\tau(g_t)$ is an homotopy of Fredholm operators from $T = \tau(z)$ to $\tau(g)$ so that $\tau(g)$ and T have the same index. Finally, from the properties of g listed above, and from the transformation law, we know that the representing functional for $\tau(g)$ is constant inside the unit disk, and its value is the same as the value at λ for the representing functional for T. Thus, it suffices to evaluate the constant for $\tau(g)$.

Thus, the problem is reduced to the case when the essential spectrum of $T = X + iY$ is the unit circle, i.e., to when T is unitary modulo the compacts. Then we know [B-D-F] that T is actually a compact perturbation of some multiple of a shift or its adjoint. However, we do not need this fact. We may proceed very naively as follows. Let T act on H. Let $n = \text{ind } T = \dim \ker T - \dim \text{coker } T$. We take an auxiliary space L on which we have an operator S which is an isometry of multiplicity n (or co-isometry if n is negative). We put $H' = H \oplus L$ and consider on H' the operator $T' = T \oplus S$. Then T' is again unitary modulo the compacts, and is of Fredholm index zero. Moreover, $[T'^*, T']$ is still trace class. If ℓ and m are the representing functionals for T and S, then evidently the representing functionals for T' is $\ell + m$. Now we know from §1 that m is just $\frac{n}{2\pi i}\, dxdy$ on the unit disk. Thus, if we can show $\ell + m = 0$, then $\ell = (\frac{-n}{2\pi i})\, dxdy$ on the unit disk as desired. We now do this.

<u>Proposition 7.2</u>: Suppose T is unitary modulo the compacts and of Fredholm index zero, and that $[T^*, T]$ is trace class. Then the bilinear form $\langle\ ,\ \rangle$ on $P\mathfrak{A}(T, T^*)$ is identically zero. In particular $\text{tr}([T^*, T]) = 0$.

<u>Proof</u>: What we must show is that $\text{tr}([T^{*m}, T^n]) = 0$ for all m and n. Any trace class perturbation of T will still satisfy the hypotheses of the proposition and it is clear (Proposition 3.4) that perturbing T by a trace class operator will not affect $\text{tr}([T^{*m}, T^n])$. Since T is Fredholm, of index zero, we may perturb T by a finite rank operator so that T is invertible.

So take T invertible and consider the algebra generated by T, T^*, T^{-1} and T^{*-1}. This is a *-algebra and since $[T^*, T^{-1}] = T^{-1}[T, T^*]T^{-1}$ and so forth, the commutator ideal of this algebra consists of trace class operators. By Corollary 2 in §4, the form $\langle R, S\rangle = \text{tr}([R, S])$ on this algebra depends only on the images of R, S modulo the compacts. But under our hypotheses $T^* - T^{-1}$ is compact, so we may compute $\text{tr}([T^{*m}, T^n]) = \text{tr}([T^{-m}, T^n]) = \text{tr}(0) = 0$ and the proposition is proved.

The proposition has the following amusing consequence, which we note explicitly.

Corollary: 'Suppose T is unitary modulo the compacts and $[T^*,T]$ is trace class. Then $tr([T^*,T])$ is minus the Fredholm index of T.

This completes the proof of points ii) and iii) of the main theorem.

Remark: It seems that there should be an analogue of the above proof that would proceed more directly, avoiding the necessity of reducing the general case to the case of an unitary modulo the compacts. The idea would be to use the mapping $z \to 1/z$ to show directly that if T were invertible, then its representing functional is zero near zero. Such a proof would probably have the advantage of freeing the proof entirely from the Hilbert space, C^*-algebra context, making the same general results valid for an a.c. pair with C^n-(separate) functional calculi.

§8. Smoothness of $g(\mu,\nu)$; connections with other work

The foregoing theory is related to other current work in operator theory. We have in mind especially Pincus $[P_1]$, $[P_2]$, $[P_4]$, Berger-Shaw [B-S], and Putnam $[Put_1]$, $[Put_2]$. We have already mentioned some of these connections. Here we will prove some propositions and raise some questions that hopefully ties this work together.

We are as always talking about an a.c. pair X,Y. We take for granted now that the representing functional for $\langle\ ,\ \rangle$ on $\mathcal{PO}(X,Y)$ is a planar measure, and we call it dP. The connections with other work come mostly when one asks questions about the fine structure of dP, mostly about the smoothness of dP, and its relation to other properties of $T = X + iY$. Thus, we will organize our discussion around a sequence of such questions.

Question 1: Is dP always absolutely continuous with respect to Lebesgue measure?

Of course, we know the answer is yes off $e\sigma(T)$. If E is very thin, e.g. a union of smooth curves, then as we will explain later (see also Proposition 7.2, corollary) dP is determined by the integers it determines on $\mathbb{C} - e\sigma(T)$. Thus, for thin $e\sigma(T)$ the answer is also yes. As we have said, Pincus has shown that his $g(\mu,\nu)$ function is $2\pi i \frac{dP}{dxdy}$, so when he can construct $g(\mu,\nu)$ the answer is also yes. This construction depends on some weak technical assumptions. If T is seminormal, his construction is valid, so the answer is yes then. Also, when the commutator has finite rank a nice estimate exists. We give a new proof of it.

<u>Proposition 8.1</u>: If $[X,Y]$ has rank n, then

$$\left| \frac{dP}{dxdy} \right| \leq \frac{n}{2} .$$

<u>Proof</u>: The proof is extremely elementary. We recall from the definition of dP that the value of dP on a rectangle $[a,b] \times [c,d] = I_1 \times I_2$ is the trace of $[X_1,Y_2]$ where X_1 and Y_2 are obtained by "cutting down" X and Y. Specifically, let P be the spectral projection of X for the interval $[a,b]$. Put $X' = PX$ and $Y' = PYP$. Let Q be the spectral projection of Y' for the interval $[c,d]$. Then $X_1 = QX'Q$ and $Y_2 = QY'$. Now we see $\sigma(X') \subseteq [a,b]$, so $\|X' - \frac{a+b}{2}\| \leq \frac{|b-a|}{2}$. Therefore, $\|X_1 - \frac{a+b}{2}\| < \frac{|b-a|}{2}$. Similarly $\sigma(Y_2) \subseteq [c,d]$, so $\|Y_2 - \frac{c+d}{2}\| \leq \frac{|c-d|}{2}$. Thus,

$\|[X_1,Y_2]\| = \|[X_1 - \frac{a+b}{2}, Y_2 - \frac{c+d}{2}]\| \leq 2\frac{|b-a||c-d|}{4}$. We also compute $[X_1,Y_2] =$

$$QX'QY' - QY'QX'Q = Q[X',Y']Q = Q(PXPYP - PYPX)Q = QP[X,Y]PQ .$$

Hence rank $([X_1,Y_2]) \leq$ rank $[X,Y] = n$. Therefore,

$$\operatorname{tr}([X_1,Y_2]) \leq \|[X_1,Y_2]\|_1 \leq n\|[X_1,Y_2]\| \leq (\tfrac{n}{2})|b-a||c-d| .$$

Since this holds for all rectangles $I_1 \times I_2$, the proposition follows

We will give an example later which shows that this estimate can be very bad.

For the rest of this section when dP is absolutely continuous we will denote the Randon-Nikodym derivative $2\pi i \frac{dP}{dxdy}$ by $g(x,y)$.

There are some indications that the size of $g(x,y)$ is related to spectral multiplicity. Along these lines we ask

Question 2: If $T = X+iY$ is hyponormal, then is $g(x,y)$ bounded by the spectral multiplicity of X (or Y)?

This is verified in simple cases, and again it can be a bad estimate of $\sigma(T)$ has a complicated shape. See [P_1] , [H] . A positive answer would give a strengthening of the main theorem of [Put_2] . Conversely, a positive answer to the following question would allow one to deduce Question 2 from [Put_1] .

Question 3: If $T = X + iY$ is hyponormal and $E_{[a,b]}$ is the spectral projection for X corresponding to the interval [a,b] , then is the spectral multiplicity of $E_{[a,b]}YE_{[a,b]}$ at most the spectral multiplicity of Y ?

A similar pair of questions serves to connect $g(x,y)$ with the work of Berger-Shaw [B-S] . Recall an operator T is said to be n-cyclic if there are n vectors $\{v_i\}_{i=1}^n$ such that the span of the rational functions in T applied to the v_i is dense in H .

Question 4: If $T = X + iY$ is hyponormal and n-cyclic, then do we have $g(x,y) \leq n$?

A positive answer to this question would strengthen the main inequality of [B-S] . We note that if the essential spectrum of T is thin, then the answer is yes, because then $g(x,y)$ just gives (minus) the Fredholm index, which is less than or equal to the codimension of $(T-(x+iy))(H)$; and it is not hard to see that a cyclic set for T must span $H \ominus (T-z)(H)$ for all $z \in \mathbb{C}$. On the other side, the inequality of [B-S] plus a positive answer to the following question would also show question 4 had a yes answer.

Question 5: Let T be hyponormal, and let $E_{[a,b]}$ be a spectral projection for X corresponding to the interval $[a,b]$. Then if T is n-cyclic, is $E_{[a,b]}TE_{[a,b]}$ also n-cyclic?

Thus the inequalities of Putnam and Berger-Shaw are naturally related to questions about the size of $g(x,y)$. A slightly different direction of inquiry concerns the relation of the (essential support of $g(x,y)$ (e.g., the support of dP) and the spectrum of T. We know already that g is supported on $\sigma(T)$. However, it is easy to see that in general the inclusion is proper. For example, if T is normal then $g(x,y) = 0$ identically. Moreover, g is invariant under trace class perturbation but $\sigma(T)$ is not. Thus the direct sum of the shift with its adjoint is a one-dimensional perturbation of the bilateral shift, so its g is zero; but its spectrum is the whole unit disk. However, in the case of hyponormal T the situation is more rigid, and the following result holds. This was conjectured and proved by Pincus $[P_4]$, and by Clancey in the case of rank-one commutator.

Proposition 8.2 (Pincus-Clancey). If $T = X + iY$ is hyponormal and completely non-normal (i.e., there exist no reducing subspaces fo T on which T is normal); then the closed support of dP is precisely $\sigma(T)$.

Proof: Let $z = x + iy$ be a point outside the support of dP. Let $[a,b] \times [c,d]$ be a small rectangle centered at z and disjoint from the support of dP. Let E be the spectral projection of X corresponding to $[a,b]$. Put $T' = X' + iY' = ETE$, acting on $E(H)$. Let F be the spectral projection for Y' corresponding to $[c,d]$. Put $T" = X" + iY" = FT'F$. By a very slight extension of the transformation law, (or the proof that dP is a measure) one sees that the representing measure $dP"$ for $T"$ is just the restriction (or cutdown) of dP to $[a,b] \times [c,d]$. Also it is very easy to show (see [H], prop. 9), that $T"$ is still completely non-normal. Finally, Putnam's spectral inclusion theorem [P] , [H], shows $\sigma(T'') = \sigma(T) \cap ([a,b]\times[c,d])$. From the choice of $[a,b]\times[c,d]$, we have $dP'' = 0$. Hence $tr([T''^*,T'']) = 0$. Since $[T''^*,T''] \geq 0$, this means it is zero, and so T'' must be normal; but since it is completely non-normal, it must be the trivial operator, and $FE(H)=\{0\}$. Thus $\sigma(T'')$ is empty, and z was not in $\sigma(T)$, which must therefore be contained in the support of dP .

Thus for the class of hyponormal, totally non-normal operators, dP and $\sigma(T)$ are related. One may ask for further relations. For example, since $g(x,y)$ is locally constant off $e\sigma(T)$, this set contains the discontinuities of $g(x,y)$. Pincus has asked whether $e\sigma(T)$ was exactly the set of discontinuities of g . It can be computed that this is true for Toeplitz operators with symbol analytic on the closed disk. However, it is false in general, as the following examples, adapted from Clancey [C_2] show.

Let S be the shift. Let $X + iY$ be the Cartesian decomposition of S . For a closed set $B \subseteq [-1,1]$, let P_B be the spectral projection of X corresponding to B . Put $S_B = P_B S P_B$, acting on $P_B H$. Then if $B = [-1,0]$ and $C = [0,1]$, we see that $S' = S_B \oplus S_C$ where $\oplus$ indicates orthogonal direct sum, again has dP equal to $\frac{1}{2\pi i}\,dxdy$ on the unit disk. One can do better. Let $B_1, B_2, \ldots$ be a sequence of disjoint Cantor sets whose union is dense in [-1.1] and whose total measure is 2 . Let $\tilde{S} = \bigoplus_{i=1}^{\infty} S_{B_i}$. Then the essential spectrum of $\tilde{S}$ is evidently the whole unit disk, but $d\tilde{P}$ is again just $\frac{1}{2\pi i}\,dxdy$. We note also that $[\tilde{S}^*,\tilde{S}]$ has infinite rank, so the estimate of Proposition 8.1 is very bad. If $\tilde{S} = \tilde{X} + i\tilde{Y}$, then the spectral multiplicity of $\tilde{Y}$ is infinite but the spectral multiplicity of $\tilde{X}$ is still one. However, if we now chop up $\tilde{S}$ in the y direction and reassemble the pieces, the result will have both real and imaginary parts with infinite spectral multiplicity, and its representing measure is still just $\frac{1}{2\pi i}\,dxdy$.

Thus, the estimate suggested by question 2 can also be very bad.

§9. Effect of compact perturbations; relation with Brown-Douglas-Fillmore

This paper has been concerned with a pair of self-adjoint operators X,Y, whose commutator is trace class. It is evident that the results are invariant under trace class perturbations. That is, if we replace X and Y by X' and Y', such that $X' - X$ and $Y' - Y$ are self-adjoint trace class operators, then X',Y' is still an a.c. pair, so our analysis applies to them. Also it is evident that the representing measures for the pairs X,Y and X',Y' are the same. More or less simultaneously with this work, a highly successful investigation of equivalence of operators under compact perturbations was conducted by Brown-Douglas-Fillmore [B-D-F]. It is natural to ask about the relation between the two lines of work.

For the case of one operator $T = X + iY$, such that $[T^*,T]$ is compact, B-D-F assign to each component of $\mathbb{C} - e\sigma(T)$ an integer, the index of $T-\lambda$ for λ in the component. Clearly the form we construct when $[T^*,T]$ is trace class produces these integers. B-D-F then go on to interpret these numbers as defining an homology class of $e\sigma(T)$. It will be seen in Part II that our form $(\ , \ )$ gives rise in a canonical way to an homology class on $e\sigma(T)$. Thus far the theories are parallel. Now, however, B-D-F show that $e\sigma(T)$ together with the homology class associated to T completely determine T up to compact perturbation. Their theory is one which is invariant under compact perturbation.

Our trace theory is not invariant under compact perturbations in the following strong sense. We say that the a.c. pair X',Y' is a compact perturbation of the a.c. pairs X,Y if $X - X'$ and $Y - Y'$ are compact. In many cases 'compactly equivalent' pairs induce forms $(\ , \ )$ and $(\ , \ )'$ on $\mathcal{S}(\mathbb{R}^2)$, which are not equal. For example, consider the operator T which is the direct sum of the shift and multiplication by z on $L^2(D)$, where D is the unit disk in $\mathbb{C}$. Then Deddens and Stampfli [Dd-S] have shown that a compact perturbation T' of T is actually normal. But dP for T is $\frac{1}{2\pi i}\, dxdy$ on D , while dP' for T' is clearly zero. We can parlay this example into an amusing observation that shows virtually any algebra, trivial (in the sense of B-D-F) , non-trivial, or otherwise contains a.c. pairs with dP not identically zero. More precisely,

<u>Proposition 9.1</u>: Let $\mathfrak{A} \subseteq L(H)$ be a C*-algebra, and suppose $LC(H) \subseteq \mathfrak{A}$. Then $\mathfrak{A}$ contains a hyponormal operator whose completely hyponormal part is the shift if and only if it contains a self-adjoint operator with uncountable spectrum.

<u>Proof</u>: If $\mathfrak{A}$ contains a shift S , then the real and imaginary parts of S have spectrum [-1,1], which is uncountable. On the other hand, if $A \in \mathfrak{A}$ is self-adjoint with uncountable spectrum σ , we may select a continuous map f from σ onto the unit disk D in $\mathbb{C}$. Then f(A) is a normal operator with spectrum D . By Weyl-von Neumann-Berg (see [B]) , any two normals with spectrum D are compact perturbations of each other. Thus, by Deddens-Stampfli, a compact perturbation of f(A) is $M \oplus S$, where M is normal and S is the shift, and this is the desired operator.

We have just seen that a compact perturbation of a.c. pairs can do violent things to its representor ℓ . There are, however, many a.c. pairs with the property that all compact perturbations of them have the same induced form (,) and consequently the same representor ℓ . We call such a pair <u>compactly stable</u>. Intuitively speaking if $e\sigma(X+iY)$ is thin the integers assigned to holes of $e\sigma(X+iY)$ will determine ℓ uniquely. These are invariant under compact perturbations. Rigorously speaking Proposition 7.2 implies that if

$e\sigma(X+iY)$ is the circle then X,Y are compactly stable, and this result can be used to show that if $e\sigma(X+iY)$ is a finite union of smooth arcs then X,Y are compactly stable. Note that the operator in the Deddens-Stampfli example has very thick essential spectrum. Thus, we pose

Question 1: Is an a.c. pair X,Y compactly stable if and only if its joint essential spectrum has Lebesgue measure equal to zero? Note that the answer to this question is yes if the answer to Question 8.1 is yes.

Let us summarize. Off of $e\sigma(T)$ our theory is rigid and behaves homotopy theoretically like that of B-D-F . On the essential spectrum we have a measure and our theory is measure theoretic more like classical spectral theory. Off the essential spectrum our theory is invariant under compact perturbations; on thick essential spectrum it varies wildly with compact perturbations. The theory is always invariant under trace class perturbations.

We conclude this section by emphasizing an obvious but intriguing formal fact. Suppose that X,Y are self-adjoint operators with $[X,Y]$ compact. Let $\{U_j\}$ be the connected components of $e\sigma(X+iY)$ complement and let $k_j = \frac{1}{2\pi i}$ index $[X+iY - (x_j+iy_j)I]$ for (x_j,y_j) in U_j . If $\sum_{j=1}^{\infty} |k_j|$ area $(U_j) < \infty$, then we call X,Y a <u>renormalizable pair</u>. Even though the trace of a commutator from $P\mathfrak{A}(X,Y)$ is not defined our representation theorem makes one believe that

$$t(p,q) = \sum_j k_j \int_{U_j} J(f,g)dxdy$$

would be quite a good number to think of as being the trace of $[p(X,Y),q(X,Y)]$. Quite possibly if one needed to know the 'trace' of these commutators for a particular purpose $t(p,q)$ would suffice. Assume at this point that the answer to Question 8.1 is yes; if it is not, the following comment still holds under more restrictive conditions. If a compact perturbation X', Y' of X,Y is almost commuting, our theory applies to X',Y' and $\mathrm{tr}[p(X',Y'),q(X',Y')]$ actually is equal to $t(p,q)$. Thus we might think of X',Y' as a 'renormalization' of X,Y and the 'formal trace' $t(p,q)$ is independent of which renormalization one chooses.

Larry Brown observed that any renormalizable pair X,Y has a renormalization. The procedure is to build operators T_j (using the construction in [B-S]) which have $e\sigma(T_j)$ = boundary U_j and index $(T_j-\lambda_j) = 2\pi i k_j$ for λ_j in U_j. Let T_e be a normal operator with $e\sigma(T_e) = e\sigma(X+iY)$. The orthogonal direct sum of T_e and the T_j's is an operator $T' = X' + iY'$ with $e\sigma(X+iY) = e\sigma(X'+iY')$ and with $k_i = k_i'$. By [B-D-F] the pairs X,Y and X',Y' generate the same B-D-F extension. Thus X',Y' is a compact perturbation of X,Y.

We conclude by mentioning one use for the 'formal trace'. It is a strengthening of Proposition 5.3 having the obvious proof.

<u>Proposition 9.2</u>: If X,Y are a renormalizable pair and if the symbol $\tilde{u}$ of an operator $U \in \mathfrak{A}(X,Y)$ has an extension to a function $u \in C^4(\mathbb{R}^2)$ with modulus = 1 on $e\sigma(T)$, then index $U = t(u,\bar{u})$.

§10. A determinant theory

The trace theory we have developed is given in terms of a bilinear form on a linear space of functions; the determinant theory is given in terms of a bi-multiplicative map on a group of invertible functions. We began this development with the relevant operator theoretic facts about determinants.

This approach is set on a group G of invertible operators with the property that for $A, B \in G$ their 'multiplicative commutator' $\{A,B\} = ABA^{-1}B^{-1}$ is the identity plus a trace class operator. Such a group will be called an almost commuting group. Let Λ denote the normal subgroup of G which consists of operators of the form

1 + trace class. Recall that any operator in Λ will have a determinant and it is given by $\det(1+K)$ = product of the eigenvalues of $I + K$ (counting multiplicity). See [G-K] for a detailed discussion of such determinants. Define a map on G by

$$\Delta(A,B) = \det\{A,B\}$$

for A,B in G. It is the obvious analog of the trace form. The basic properties of δ are

1. $\Delta(A,B) = \Delta(B,A)^{-1}$,
2. $\Delta(\lambda,A) = 1$ if $\lambda \in \Lambda$
3. $\Delta(AB,C) = \Delta(A,C)\Delta(B,C)$,
4. $\Delta(A,B) = 1$ if A and B commute.

Proof of Properties: These facts follow from well-known properties of det. Fact 1 follows from $\det C = (\det C^{-1})^{-1}$. Fact 4 follows from $\det I = 1$. Fact 3 follows because $\{AB,C\} = ABCB^{-1}C^{-1}A^{-1}ACA^{-1}C^{-1} = A\{B,C\}A^{-1}\{A,C\}$ and $\det(EF) = \det(FE) = \det E \det F$. Fact 2 follows because $\det(\lambda A\lambda^{-1}A^{-1}) = \det \lambda \det(A\lambda^{-1}A^{-1}) = \det \lambda \det(\lambda^{-1}) = 1$.

A consequence of these facts is that Δ induces a bi-multiplicative map (bi-character) δ on the commutative group G/Λ . The authors suspect that an abstract study of multiplicative maps parallel to the study of linear maps given in §2 would be amusing. However, we have not pursued this. What we shall do now is describe the connection between the trace theory and the determinant theory.

Suppose as usual that X,Y is an almost commuting pair. Let $\mathfrak{A}_1$ be a maximal subalgebra of $\mathfrak{A}(X,Y)$ whose self-adjoint pairs almost commute; this is a reasonable setting to use (see §3). Let G be the group of invertible operators in $\mathfrak{A}_1$. Note that if $AB - BA = K$ trace class, then $ABA^{-1}B^{-1} = I + KA^{-1}B^{-1}$ is in Λ ; thus G is almost commuting. Now we apply the preceding to get an induced bi-character δ . Let $\tilde{M}$ denote the group of all C^∞ functions which are invertible on a big closed box containing $\sigma(T)$.

Let M denote the subgroup of functions f so that $\tau(f)$ has Fredholm index equal to zero. If $f \in M$, then some finite rank perturbation $\tau'(f)$ of $\tau(f)$ is in G. Define δ on M by

$$\delta(f,g) = \det\{\tau'(f),\tau'(g)\}$$

and note that Facts 2,3 imply that this definition does not depend on which perturbation $\tau'(f)$ of $\tau(f)$ one takes. The map δ is a bi-character on M with the property that it is 1 whenever f and g are functions of the same function. We call such maps collapsing bi-characters.

At this point we write down another collapsing bi-character on $\tilde{M}$. Let ℓ denote the representing functional for X,Y. Define for f,g in $\tilde{M}$

$$\beta(f,g) = \exp \ell\left(\frac{J(f,g)}{fg}\right) .$$

Formally, this is just

$$\beta(f,g) = \exp \operatorname{tr}[\log \tau(f) , \log \tau(g)]$$

The map β is a bi-character, since

$$\frac{1}{fhg} J(fh,g) = \frac{1}{fg} J(f,g) + \frac{1}{hg} J(h,g)$$

as one can easily compute; the collapsing property of β follows from that of the Jacobian.

The authors conjectured that $\beta(f,g) = \delta(f,g)$ and this was verified by Pincus $[P_3]$ for functions f and g which possess logarithms. His proof is included for the sake of completeness. Thus the 'trace invariant' and the 'determinant invariant' are most likely equivalent. The authors have not studied variants on this approach very much at all, and conceivably a closer look at determinants might give some essentially new information. For example one possibility exists with determinants which does not exist with traces when one takes the viewpoint of B-D-F. The family of extensions forms a group.

The group may have torsion but the trace invariant is a map into the real numbers and so loses all torsion information. The determinant invariant is a map into the multiplicative complexes and consequently might not suppress the torsion information. Another possible advantage of determinants and of the multiplicative commutator is that $\{A,B\}$ may be I + trace class even when A and B are unbounded. Thus the determinant invariant possibly will be easily defined in some physical situations when the trace invariant will be difficult to define. Thus it seems that though the two theories will be mostly the same one will more useful for some situations and the other for other situations.

Proposition 10.1: If f and g are $C^\infty(\mathbb{R}^2)$ functions, then

$$\det\{\tau'(e^f),\tau'(e^g)\} = \exp \int J(f,g)dP$$

Proof: The proof relies heavily on the Baker-Campbell-Hausdorff expansion which says $e^W e^Q = e^{m(W,Q)}$ where

$$m(W,Q) = W + Q + \frac{1}{2}[W,Q] + \frac{1}{12}[W,[W,Q]] + \ldots$$

provided that W and Q have sufficiently small norm. Suppose that $[W,Q]$ is trace class. Since $e^W e^Q e^{-W} e^{-Q} = \exp m(m(W,Q),m(-W,-Q))$ $\exp([W,Q] + \text{higher commutators})$, since $\det e^C = \exp \operatorname{tr} C$, and since $\operatorname{tr}(\text{higher commutator}) = 0$ one gets

$$\det e^W e^Q e^{-W} e^{-Q} = \exp \operatorname{tr}[W,Q] \quad .$$

To extend this result to W and Q of arbitrary norm, multiply them by a parameter r and note that both sides are analytic in r. In order to complete the proof one need only show that $e^{\tau(f)} - \tau(e^f)$ is trace class. We shall be sketchy at this point and simply say that the estimates behind Proposition 3.3 (ii) give this; one has the equation $e^{\tau(f)} - \tau(e^f) = \sum_{n=0}^{\infty} \frac{1}{n!}[\tau(f)^n - \tau(f^n)]$ and those estimates on $||\tau(f)^n - \tau(f^n)||_1$ are quite strong.

Remark: In Pincus' Theory there are two important invariants associated with an almost commuting pair X, Y. He calls one the $E(\ell, z)$ function and one the $g(\mu,\nu)$ function. We have already discussed the g function. The E function is defined by

$$E(\ell,z) = 1 + \frac{1}{\pi i} K(X-\ell)^{-1}(Y-z)^{-1}K^*J$$

for $\ell \notin \sigma(X)$, $z \notin \sigma(Y)$ where $[X,Y] = \frac{1}{\pi i} K^*JK$ with $J^2 = I$ and $J = J^*$. Note that

$$\det E(\ell,z) = \det(1 + [X,Y](X-\ell)^{-1}(Y-z)^{-1})$$

$$= \Delta(X-\ell, Y-z)$$

Thus if the functions $x-\ell$ and $y-z$ are expressed as exponentials (assuming ℓ and z not real), Proposition 10.1 gives that

$$\det E(\ell,z) = \exp \int \frac{1}{x-\ell}\frac{1}{y-z}\, dP$$

which is the formula Pincus uses to define his $g(x,y) = \frac{1}{2\pi i}\frac{dP}{dxdy}$ function.

We conclude this section by mentioning that the determinant theorem for Toeplitz operators looks a bit like a classical and quite useful theorem of Szego [see ChX[M-W]]. We have no deep comments to offer, but it is amusing that they look a little bit alike.

Suppose that T_a and T_b are two index zero Toeplitz operators with C^1 generating functions having logarithms $\ell_a(\theta) = \Sigma\, a_n e^{in\theta}$ and $\ell_b(\theta) = \Sigma\, b_n e^{in\theta}$. Then Proposition 10.1 along with Stokes theorem implies

$$\det(T_aT_bT_a^{-1}T_b^{-1}) = \exp \sum_{-\infty}^{\infty} j\, a_{-j}b_j$$

The classical Szego theorem says that

$$\lim_{N\to\infty} \frac{1}{N} \det P_NT_aP_N = \mu \exp \sum_{j=1} j\, a_{-j}a_j$$

where P_N are the coordinate projections on ℓ^2 defined in §1 and $\mu = \exp a_0$. If one considers the special case $b(\theta) = a(-\theta)$, then the right side of our determinant theorem is $\sum_{-\infty}^{\infty} j\, a_{-j} a_{-j}$ and so the two theorems look a little more alike. Szego's seems to express some cross term phenomenon while ours is a bit more quadratic.

APPENDIX

The appendix is devoted to proving Proposition 1.6.1. See subsection 6 for definitions of $E_{[a,b)}$, $F_{[c,d)}$, $\mathcal{P}$ etc. Recall

$$\tilde{P}([a,b)\otimes[c,d)) = \mathrm{tr}(E_{[a,b)}F_{[c,d)}E_{[a,b)}[X,Y])$$

We begin with a lemma. Suppose E_i is a finite sequence of spectral projections for X corresponding to a sequence of disjoint left closed, right open intervals on the line. Let $\mathcal{B}$ denote the Borel Functions on $\mathbb{R}$ which have compact support.

<u>Lemma A.1</u>: If a_i is a sequence of real numbers and h is in $\mathcal{B}$, then

$$\mathrm{tr}(\sum_i^N a_iE_i\; h(\sum_i^N a_iE_iY \sum_i^N a_iE_i) \sum_i^N a_iE_i\; [X,Y]) =$$

$$= \sum_i^N a_i^2\; \mathrm{tr}(E_i\; h(a_i^2E_iY\; E_i)\; E_i[X,Y])$$

<u>Proof</u>: First observe that if w is a $\mathcal{B}$ function

$$w(X)Y\, w(X)\; X - X\; w(X)\; Y\, w(X) = -w(X)[X,Y]w(X)$$

One can now multiply both sides of this by $k(g(X)Yg(X))$ where $k \in \mathcal{B}$ and obtain

$$\mathrm{tr}([w(X)Yw(X),\; k(w(X)Yw(X))X\; k(w(X)Yw(X)]) =$$

$$= -\;\mathrm{tr}(w(X)k(w(X)Yw(X))^2w(X)[X,Y])$$

Therefore, if w and k are $C_0^\infty(\mathbb{R})$ functions

$$\text{(A.1)}\qquad \ell(w(x)^2k(w(x)^2y)^2) = \mathrm{tr}(w(X)k(w(X)Y\; w(X))^2w(X)[X,Y])$$

Functions of the type $w(x)^2k(w(x)^2y)^2$ occur throughout what follows. We shall use them to show that $\tilde{P}$ has measure theoretic properties. Things would be simpler if we could work with functions of the form $w(x)^2k(y)^2$. However, if w and k are characteristic functions of intervals then $w(x)^2k(w(x)^2y)^2 = w(x)^2k(y)^2$; our procedure is based on this fact.

It is easy to see that if g_i is a sequence of real valued functions with disjoint support and if a_i is a sequence of real numbers, then

$$\text{(A.2)} \qquad \Sigma\, a_i^2 g_i(x)^2 h(\Sigma\, a_j^2 g_j(x)^2 y) = \Sigma\, a_i^2 g_i(x)^2 h(a_i^2 g_i(x)^2 y) \quad .$$

Suppose that g_i and h are $C_0^\infty(\mathbb{R})$ functions, apply ℓ to both sides of the equation and then use (A.1) to obtain

$$\text{(A.3)} \quad \operatorname{tr}(\Sigma\, a_i g_i(X) h(\Sigma\, a_i g_i(X) Y\, \Sigma a_i g_i(X))\, \Sigma\, a_i g_i(X)[X,Y]) =$$

$$= \Sigma\, a_i^2 \operatorname{tr}(g_i(X) h(a_i^2 g_i(X) Y\, g_i(X)) g_i(X)[X,Y])$$

Note that this formula holds whenever g and h are Borel functions. In order to finish the lemma let the g_i's approximate the characteristic function of the ith interval in the obvious manner, and then observe that a similar approximating argument allows us to take h to be any $\mathcal{B}$ function. Here we have used the facts that if self-adjoint $A_n \to A$ in the strong operator topology $h(A_n) \to h(A)$ and $\operatorname{tr}(A_n C) \to \operatorname{tr}(AC)$ provided that C is trace class. End of Proof.

We now prove that $\tilde{P}$ is a finitely additive set function on the field of sets Σ generated by $\mathcal{R}$. It suffices to demonstrate that if a rectangle C in $\mathcal{R}$ is the union of a finite disjoint collection $\{C_i\}$ of sets in $\mathcal{R}$ then $\tilde{P}(C) = \Sigma\, \tilde{P}(C_i)$. See for reference Ch.12 Prop. 11 and Ch. 12, sec. 2 ex. 4a of [Rn] . This will follow provided that if a rectangle C in $\mathcal{R}$ is the union of two disjoint rectangles C_1 and C_2 in $\mathcal{R}$

$$\text{(A.4)} \qquad \tilde{P}(C) = \tilde{P}(C_1) + \tilde{P}(C_2)$$

Given a rectangle C there are only two ways to split it into two rectangles: vertically or horizontally. The (vertical) decomposition of $[a,b) \otimes [c,d) = C$ into two rectangles $C_1 = [a,b) \otimes [c,c_1)$ and $C_2 = [a,b) \otimes [c_1,d)$ satisfies (A.4) simply by definition of $\tilde{P}$.

Consider the horizontal decomposition of C into $C_1 = [a,\xi) \otimes [c,d)$ and $C_2 = [\xi,b) \otimes [c,d)$. We shall apply lemma A.1 with h the characteristic function of $[c,d)$, with E_1 (resp. E_2) equal to the spectral projection for X on $[a,\xi)$ (resp. $[\xi,d))$, with $1 = a_1 = a_2$, and with every other quantity equal to zero. We get

$$\operatorname{tr}([E_1+E_2]h([E_1+E_2]Y[E_1+E_2])[E_1+E_2][X,Y])$$

$$= \quad \operatorname{tr}(E_1h(E_1YE_1)E_1C) + \operatorname{tr}(E_2h(E_2YE_2)E_2[X,Y])$$

that is $\tilde{P}(C) = \tilde{P}(C_1) + \tilde{P}(C_2)$. We have shown that $\tilde{P}$ is finitely additive.

Now we prove that $\tilde{P}$ has bounded total variation. By Ch. III sec. 1.4 Lemma 5 of [D-S] we need only show that the maximum of $|\tilde{P}(S)|$ over all S in Σ is bounded. Any S in Σ is the union of finitely many disjoint rectangles S_j in R . Without loss of generality we may assume that there exist finitely many disjoint intervals $[a_\ell,b_\ell)$ on the x-axis and finitely many disjoint intervals $[c_k,d_k)$ on the y-axis so that each S_j has the form $[a_\ell,b_\ell)\otimes[c_k,d_k)$. Let S^ℓ denote the union of the S_j whose x-projection is $[a_\ell,b_\ell)$ and let P_ℓ be the spectral projection for $E_{[a_\ell,b_\ell)}Y\,E_{[a_\ell,b_\ell)}$ associated with the set of y coordinates for S^ℓ . Then

$$\tilde{P}(S^\ell) = \operatorname{tr}(E_{[a_\ell,b_\ell)}P_\ell E_{[a_\ell,b_\ell)}[X,Y]) \quad .$$

Since the ranges of $E_{[a_\ell,b_\ell)}$ for distinct ℓ are orthogonal

$$\|\sum_\ell E_{[a_\ell,b_\ell)}P_\ell\, E_{[a_\ell,b_\ell)}\| \le 1 \quad \text{and consequently}$$

$$|\tilde{P}(S)| \le \operatorname{tr}(|[X,Y]|) \tag{A.5}$$

Remark (Larry Brown): The inequality (A.4) and III §1.4 Lemma 5 of

[D-S] only give that the total variation of P is less than or equal to $2\,\mathrm{tr}|[X,Y]|$. One can get $\mathrm{tr}|[X,Y]|$ as the bound, since the projections over the rectangles S_j can be weighted with either a + or - sign without effecting the estimator.

Remark: The general operator theoretic fact at work here is: If E_i are a sequence of mutually orthogonal projections, then $\lambda: \mathcal{T} \to \mathcal{T}$ defined by

$$\lambda(T) = \Sigma\, E_i\, T\, E_i$$

is a contraction in the trace norm. This follows because λ is clearly a norm contraction on the $\mathcal{LC}(\mathcal{H})$ and $\mathcal{T}$ is the dual of $\mathcal{LC}(\mathcal{H})$; thus λ adjoint is a trace norm contraction on $\mathcal{T}$. However, $\mathrm{tr}(\lambda(C)T) = \mathrm{tr}(C\lambda(T))$ and so λ itself is a trace norm contraction.

Now we prove that $\tilde{P}$ is regular on the field Σ , generated by $\mathcal{R}$. Define $f(r,s) = \tilde{P}([-||X||r,)\otimes[-||Y||,s))$. Since $\tilde{P}$ has finite total variation it is a straightforward though tedious exercise in measure theory to show that $\tilde{P}$ is regular if and only if f is separately left continuous in r and s . Thus, what we prove is that f is separately continuous. Let $a = -||X||$ and $c = -||Y||$.

The definition of $\tilde{P}$ gives

$$f(r,s) - f(r,t) = \mathrm{tr}(F_{[s,t)}[X,Y]F_{[s,t)})$$

As $t \to s$ the projections $F_{[s,t)} \to Q$ the projection onto the eigenspace for $E_{[a,r)}Y\,E_{[a,r)}$ with eigenvalue s . Since $Q[X,Y]Q = Q\,E_{[a,r)}[X,Y]E_{[a,r)}Q = QXQs - sQXQ = 0$, we get that as $t \to s, f(r,t) \to f(r,s)$. As $s \to t$ the projections $F_{[s,t)} \to 0$ and so $f(r,s) \to f(r,t)$. Thus, we have continuity in the second variable.

Now we examine the first variable in f . Let $B(q,r) = {}$ $= E_{[q,r)}[X,Y]E_{[q,r)}$. An argument like the one above implies that $\mathrm{tr}|B(q,r)| \to 0$ as $r \to q$ or as $q \to r$. Since $\tilde{P}$ is additive

$$f(r,s) - f(q,s) = \tilde{P}([q,r)\otimes[c,s)) .$$

If $F_{[q,r) \otimes [c,s)}$ denotes the spectral projection of $E_{[q,r)} Y E_{[q,r)}$ associated with $[c,s)$, then

$$|f(r,s) - f(q,s)| = |tr(F_{[q,r) \otimes [c,s)} B(q,r))|$$
$$\leq tr|B(q,r)|$$

Thus, f is separately continuous.

The continuity of f implies more than just P is regular; one gets that the P measure of any vertical or horizontal line is zero. The transformation law of §7 allows one to conclude that the P measure of smooth curves is zero.

The final thing to prove is that $\ell(f) = \int f dP$. This is a bit tedious but straightforward. In some sense it follows by doing the process we have just finished in reverse, namely, in extending formulas which hold for characteristic functions to ones which hold for smooth functions.

Part II

ALGEBRAS OF HIGHER DIMENSIONAL SINGULAR INTEGRAL OPERATORS

In Part I we developed a 'trace theory' for an elegant and simple abstraction of a pair of operators - one a singular integral operator on the line, the other a multiplication operator. The first thing we do in Part II is describe how these results extend to families of such operators. Next we turn to higher dimensional singular integral operators (e.g., pseudo-differential operators of order zero on a manifold). A suitable formal and rather algebraic definition for these operators exists and could be thought of loosely as 'almost solvable' instead of almost commuting which characterizes the one-dimensional case. Rather surprisingly, one can define a 'trace invariant' for families of these 'algebraic' singular integral operators and the authors believe that a prolonged study of it and of the 'algebraic' singular integral operators themselves will be rewarding. In our work we have concentrated on the two lowest dimensions and this part (except for §1) announces our results on dimension 2 (Part I concerned dimension 1).

One might suppose that what we have described is notationally involved and complicated. In fact, the basic formulation is quite clean. The representation theorem in Part I can be stated in terms of differential forms and its more general analogues have elegant statements in these terms. The most striking fact, however, is that the index results in Part I (Main Theorem, part (ii) and (iii)) are basically homological results and carry over in that form to the general situation.

One of the main results of Part I is an index theorem. The homological reformulation we give in this part suggests in a fairly natural way (especially with the hindsight given by B-D-F) a lengthy general program, the culmination of which would be an abstract and extremely general index theorem. Section 3 is devoted to formulating an approach to this goal and to describing our progress so far. We can establish a one-dimensional index result in complete generality. We know what the answer is in the two-dimensional case, but one substantial gap remains in the proof.

§1. A different view of almost commuting pairs

The best way to state the main theorem in Part I is by using differential forms. The index result, parts (ii) and (iii) of the main theorem are actually statements about homology classes (DeRham). The key observation is that if $f, g \in S(R^2)$, then $df \wedge dg = J(f,g)dx \wedge dy$. Thus, we may think of integration of $J(f,g)$ against a measure as a linear functional applied to the closed form $df \wedge dg$. Thus, it is not hard to believe that the representation theorem (Wallach's Lemma) says: there exists a linear functional ℓ defined on the smooth exterior two forms on $\mathbb{R}^2$ such that $(p,q) = tr(p(X,Y),q(X,Y)) = \ell(dp \wedge dq)$. Now parts (i), (ii), (iii) and (iv) of the theorem give us properties of ℓ . The most important of these properties are the index results (ii) and (iii). Part (ii) says that the linear functional ℓ vanishes on derivatives of one-forms supported on the complement of $e\sigma(T)$. This property allows one to use a straightforward canonical construction to associate with ℓ a second homology class h for $\mathbb{R}^2$ relative to $e\sigma(T)$. Part (iii) of the theorem then identifies this class and shows in particular that it is integral. According to Proposition 5.3, the homology class h has the property that if the operator U in the C*algebra generated by X and Y has an extended symbol $u \in S(R^2)$ whose modulus is one on $e\sigma(T)$, then

$$\text{index } U = h(du \wedge d\bar{u}) .$$

Before stating all of this formally, we introduce our notation and conventions. The best language for a perfectly precise description of our results is sheaf theory, and it will be used in the detailed paper. However, one can get an extremely good understanding of the results without it and we avoid sheaf theory in our present discussion. We shall be thinking of differential forms as if they were test functions in distribution theory and studying linear functionals on them. Such objects are called 'currents' and their theory has been extensively developed (see [F]), but for our purposes one needs to know essentially nothing about them. Suppose that Ω is an open set in $\mathbb{R}^n$. We will be using C^∞-differential forms with compact support in Ω. Recall that any differential n-form v on $\mathbb{R}^n$ can be written as

$$v = \sum_{i_1 < i_2 < \ldots i_k} f_{i_1 i_2 \ldots i_k} dx_{i_1} \wedge \ldots \wedge dx_{i_k}$$

where $f_{i_1 \ldots i_k}$ is in $C^\infty(\mathbb{R}^n)$. The form v is said to have compact support in Ω provided that its coefficients have compact support in Ω, that is, belong to $C_o^\infty(\Omega)$. The set of $\Lambda^k(\Omega)$ of k-forms on Ω is a locally convex linear topological space in the topology acquired from $\mathcal{D} = C_o^\infty(\Omega)$, see [Ru]. Let $C\Lambda^k(\Omega)$ denote the closed and $E\Lambda^k(\Omega)$ denote the exact k-forms on $\mathbb{R}^n$ with coefficients in $C_o^\infty(\Omega)$. If E is a closed subset of Ω one lets $H_k(\Omega, E)$ denote the k-th homology group of Ω relative to E. Let us amplify this last definition a bit. The k-th relative cohomology class $H^k(\Omega, E)$ is just the space of all closed k-forms which vanish near E modulo d of the k-1 forms that vanish near E.

The homology space $H_k(\Omega, E)$ is just the dual of $H^k(\Omega, E)$ and so a linear functional ℓ on $C\Lambda^k(\Omega)$ which vanishes on $E\Lambda^k(\Omega{\sim}E)$ induces an element h_ℓ in $H_k(\Omega, E)$. Note that h_ℓ need not necessarily determine ℓ. That is, h_ℓ only depends on the behaviour of ℓ away from E.

We will soon state the main theorem of Part I in this language. However, we will do it in greater generality, namely, for almost commuting algebras. We say that an algebra $\mathfrak{A}$ of operators is <u>almost commuting</u> (a.e.), provided that the commutator of any two operators in the algebra is trace class. Suppose that $X_1,\ldots,X_n$ are self-adjoint operators so that any pair of them is almost commuting; we call this an <u>almost commuting</u> family. Then $P\mathfrak{A}(X_1,\ldots,X_n)$ the polynomial algebra generated by $X_1,\ldots,X_n$ is almost commuting (abbreviate it by $P\mathfrak{A}$). We proceed exactly as in Part I. Define for p,q in P_n, the commutative algebra of all polynomials on R^n, a bilinear form

$$(p,q) = \mathrm{tr}[p(X_1,\ldots,X_n),q(X_1,\ldots,X_n)]$$

where $p(X_1,\ldots,X_n)$ and $q(X_1,\ldots,X_n)$ are operators in $P\mathfrak{A}$ gotten by substituting $X_1,\ldots,X_n$ formally into p and q in any order whatsoever. The estimates in §3 carry over directly to this case and so we may consider $(\; , \;)$ to be a bilinear form on $S(R^n)$ with with compact support. Let $E = e\sigma(X_1,\ldots,X_n)$ denote the natural embedding into $\mathbb{R}^n$ of the maximal ideal space of the C*-algebra $C^*\mathfrak{A}$ generated by $X_1,\ldots,X_n$ modulo its intersection with the compacts $LC(H)$. E is called the joint essential spectrum of $X_1,\ldots,X_n$. If $A \in C^*\mathfrak{A}$ its <u>symbol</u> is the function on E arising from A in the Gelfand representation of $C^*\mathfrak{A}/C^*\mathfrak{A} \cap LC(H)$. Our representation theorem is

<u>Theorem 1.</u> If $X_1,\ldots,X_n$ is an almost commuting family of operators, then there exists a continuous linear functional ℓ on $C\Lambda^2(\mathbb{R}^n)$ having compact support such that

$$(p,q) = \mathrm{tr}[p(X_1,\ldots,X_n),q(X_1,\ldots,X_n)] = \ell(dp \wedge dq)$$

for $p,q \in C_o^\infty(\mathbb{R}^n)$. The linear functional ℓ has the properties

(i) ℓ is identically zero on $E\Lambda^1(\mathbb{R}^n \sim E)$.

(ii) If U is in the C*algebra generated by $X_1,\ldots,X_n$, and if the symbol $\tilde{u}$ of U has an extension to a function u in $C^\infty(\mathbb{R}^n)$, then the homology class h_ℓ induced by ℓ canonically determines its index, namely,

$$\text{index } U = h_\ell(du \wedge d\bar{u}) .$$

The proof of this is very much like the proof in Part I. Everything has to be carried over to n variables and though this is not trivial it works out very well. Also we should observe that the type of information specified by h_ℓ is first homological information about E . The relative 2-class h_ℓ gives rise to an absolute 1-class of E via the boundary isomorphism $H^1(E) \approx H^2(R^n,E)$. The boundary isomorphism essentially amounts to Stokes Theorem.

The index theorem extends perfectly to matrix valued functions which we now describe. The notation will be as follows. If $\mathfrak{A}$ is an operator algebra, let $M_m \otimes \mathfrak{A}$ denote the $m \times m$ matrices with entries which are operators in $\mathfrak{A}$, let $CM_m(E)$ denote the continuous $m \times m$ matrix valued functions on E , and let $\Sigma_m(\mathbb{R}^n)$ denote the $m \times m$ matrix valued functions on R^n with entries in $\mathcal{S}(\mathbb{R}^n)$. The elements of $M_m \otimes \mathfrak{A}$ are considered to be operators on $H \oplus H \oplus \ldots \oplus H$. If $X_1,\ldots,X_n$ is a family of self-adjoint operators which is commutative modulo the compacts and $\mathfrak{A}$ is its C*-algebra, then the symbol of an operator W in $M_m \otimes C^*\mathfrak{A}$ is the matrix function in $CM_m(E)$ whose entries are the symbols of the entries of W . The index theorem for a matrix of operators is

Theorem 2. Suppose that $X_1,\ldots,X_n$ is an almost commuting family and $\mathfrak{A}$ is the C*-algebra it generates. Then the index of an operator U in $M_m \otimes \mathfrak{A}$ with unitary valued symbol having an extension to a matrix function $u = \{u_{ij}\}$ in $\Sigma_m(\mathbb{R}^n)$ is

$$\text{index } U = \sum_{j,k=1}^{n} h_\ell(du_{jk} \wedge d\bar{u}_{jk})$$

Remark: The difference between Theorem 1 and Theorem 2 may be thought of as follows. Theorem 1 says our homology class is the one-dimensional component of the index class [in the sense of B-D-F] for $\mathfrak{A}$. Theorem 2 says all the higher components of the index class are all zero, so our class is the entire index class. Thus Theorem 2 is the complete index theorem for almost commuting algebras. It also forms an absolute analytical obstruction to forming almost commuting algebras: no algebra with higher index invariants can be generated by an almost commuting family. Thus, for example, if an algebra on S^3 is generated by almost-commuting operators, it is trivial in the sense of B-D-F.

§2. Higher dimensional singular Integral operators and Higher Trace Forms

In the preceding section we had an almost commuting algebra with joint essential spectrum E . We associated canonically with the algebra and a distinguished set of generators what amounted to a first homology class of E . The primary example of such a setup is an algebra of singular integral and multiplication operators on $\mathbb{R}$ or T . There are two obvious questions. Can we by some similar process associate with the algebra higher homology classes of E ? Can we study algebras of singular integral operators on $\mathbb{R}^k$ (these have commutators in the Schatten p-classes but not in the trace class)? These questions turn out to be strongly related and they occupy the rest of the paper. We approach the second one first and begin by introducing axioms on an operator algebra which abstracts the notion of a higher dimensional singular integral operator in much the same way that 'almost commuting' characterized one dimensional singular integral operators. Higher dimensional singular integral operators have the property that their iterated commutators are trace class and the abstract notion can be thought of as 'almost solvable' as opposed to 'almost commuting'. Suppose that $\mathfrak{A}_o$ is an algebra of operators in $L(H)$. We define the commutator filtration of $\mathfrak{A}_o$ to be a particular sequence $\mathfrak{A}_o \supset \mathfrak{A}_1 \supset \mathfrak{A}_2 \supset \ldots$ of ideals in $\mathfrak{A}_o$. It is the smallest sequence of decreasing ideals $\mathfrak{A}_j$ which satisfy $[\mathfrak{A}_j, \mathfrak{A}_k] \subset \mathfrak{A}_{j+k+1}$ and $\mathfrak{A}_j \mathfrak{A}_k \subset \mathfrak{A}_{j+k}$. An algebra is called almost solvable (in n-steps) if $\mathfrak{A}$ is commutative modulo the compact operators and if the ideal $\mathfrak{A}_{n+1}$ is contained in the trace class operators. Before going on, let us say a few more words about the commutator filtration. It can be characterized as follows. The subalgebra $\mathfrak{A}_1$ is the ideal generated by all commutators of operators in $\mathfrak{A}_o$, i.e., $\mathfrak{A}_1$ is the commutator ideal, the subalgebra $\mathfrak{A}_2$ is the ideal generated by products of operators in $\mathfrak{A}_1$ and by commutators of operators in $\mathfrak{A}_o$ with operators in $\mathfrak{A}_1$, the subalgebra $\mathfrak{A}_3$ is the ideal generated by products of three operators in $\mathfrak{A}_1$ and by commutators of operators in $\mathfrak{A}_o$ with operators in $\mathfrak{A}_2$, etc.

The most glaring example of an almost solvable algebra is $P_s N$ the pseudo differential operators of order zero on an n dimensional manifold N. The algebra $P_s N$ is almost solvable in n steps. Another example is TP_n the Toeplitz operators on the $2n-1$ sphere (n odd) with smooth symbol (Venugopalkrishna [V]). Here TP_n is almost solvable in n steps.

Remark. The algebras in these examples satisfy even a stronger restriction than almost solvability which we introduce now. Recall (see Ch. [R-S]) that the Schatten p-class C_p is defined to be the ideal of operators A which satisfy $\sum_{i=1}^{\infty} ||Ax_i||^p < \infty$ where x_i

is any complete orthonormal basis. An operator algebra $\mathfrak{A}$ which is almost solvable in n-steps will be called <u>tapered</u> if $\mathfrak{A}_\ell \subseteq C_{(n+1)/\ell}$, $\ell \leq n$. At this point the authors are not sure whether the stronger or weaker notion will predominate the development of the higher commutator trace theory. However, the basic definitions require only the weaker.

Now we begin to study the multilinear forms which give rise to higher homology classes. The bilinear form we have been studying is trace of the commutator. We define higher commutators in an obvious way. Define the antisymmetrization $[A_1,\ldots,A_\ell]$ of operators $A_1,\ldots,A_\ell$ on H to be the antisymmetric sum of the permuted products of A_1 through A_ℓ. For example, $[A_1,A_2]$ is just the ordinary commutator while $[A,B,C] = ABC - BAC + BCA - CBA + CAB - ACB$.

We will be studying the multilinear forms, $\mathrm{tr}[A_1,\ldots,A_\ell]$ on an almost solvable algebra. In order to show that there is an interesting theory for these forms we need three things:

1. $[\ ,\ ,\ ]$ must be trace class
2. $\mathrm{tr}[\ ,\ ,\ ]$ must induce a form on $\mathfrak{A}/\mathfrak{A}_1$

That is, $\mathfrak{A}_1$ must be in the radical of $\mathrm{tr}[\ ,\ ,\ ]$.

3. $\mathrm{tr}[\ ,\ ,\ ]$ must not be identically zero.

The third condition may seem facetious, but it is inserted to foreshadow something which happens frequently. Now let's see when these conditions are satisfied.

Suppose that $\mathfrak{A}$ is an operator algebra which is almost solvable in k-steps. If the operator $[A_1,\ldots,A_\ell]$ is trace class for any $A_1,\ldots,A_\ell$ we say that $\mathfrak{A}$ is ℓ-summable. An algebraic computation shows that any k-step almost solvable algebra is m-summable if $m \geq 2(k+1)$. However, for the interesting examples ($\mathcal{P}_s N$ and $\mathcal{TP}_n$) something stronger holds. When one of them is k-step almost solvable it is 2k-summable. This is a crucial property for us. We will call an algebra which is k-step almost solvable and 2k-summable a crypto*-(singular) integral algebra (of dimension k).

Proposition 2.1 If $\mathfrak{A}$ is a *-closed algebra which is almost solvable in k-steps and if $\mathfrak{A}$ is ℓ-summable with $\ell \geq 2k-1$, then the commutator ideal $\mathfrak{A}_1$ of $\mathfrak{A}$ is in the radical of $\langle\ ,\ \rangle_\ell$ provided that ℓ is even.

Thus, for a very reasonable class of operator algebras and for large enough even ℓ the form $\langle\ \rangle_\ell$ induces a form $(\ \ )_\ell$ on the commutative algebra $\mathfrak{A}/\mathfrak{A}_1$, and we have the prospect of developing a higher dimensional analogue to Part I.

The most glaring oddity about this is that it doesn't work unless ℓ is even. This phenomenon is fundamental and not merely technical. It is a manifestation of a finite dimensional fact. Suppose that H is finite dimensional; then $\langle\ ,\ ,\ \rangle_\ell$ is identically zero when ℓ is even and not identically zero when ℓ is odd. It is also quite reasonable from a topologist's point of view since Bott periodicity plus [B-D-F] implies that only the even relative homology classes for E are important anyway.

Next we find that most of the even forms $\langle\ \rangle_{2k}$ are identically zero, something that has already been hinted at.

Proposition 2.2 Suppose that $\mathfrak{A}$ is a *-closed algebra which is solvable in k-steps. Then for $\ell > k$, the form $\langle\ ,\ ,\rangle_{2\ell}$ is identically zero. (Note $\mathfrak{A}$ is automatically 2ℓ-summable if $\ell > k$.)

What happens in examples and what these two propositions tell us is that for a given algebra $\mathfrak{A}$ there is one interesting trace

---

*A tip of the hat to L. Zalcman for this name.

form. The higher order trace forms are identically zero, the lower order forms are not well defined. Also we see that only the crypto-integral algebras are of interest for our purposes. Henceforth, we work only with crypto-integral algebras and we call the crucial trace forms on the algebra $\mathfrak{A}$ the <u>fundamental form</u> of $\mathfrak{A}$. Note that the one dimensional crypto-integral algebras are just the almost commuting algebras.

The next task is to get a representation for the trace invariant when $X_1, X_2, \ldots, X_n$ are self adjoint operators whose polynomial algebra $P\mathfrak{A}(X_1, \ldots, X_n)$ is a crypto-integral algebra of dimension k. Such operators are called a family of <u>algebraic integral operators</u> of dimension k. We have a representation for the trace invariant $\langle\, ,\, ,\, \rangle_{2k}$ when $k = 1$, or 2; for higher k we have not and this remains an extremely intriguing open problem. Now we consider a family $X_1, \ldots, X_n$ of algebraic integral operators (dim 2) and the fundamental form $\langle\;\rangle_4$ on its polynomial algebra. We can perform estimates like those in §3 but a good bit more complicated which allow one to extend $\langle\;\rangle_4$ continuously to the $C_0^\infty(\mathbb{R}^n)$ or $S(\mathbb{R}^n)$ functions, and which allow us to construct a functional calculus $\tau : C_0^\infty(\mathbb{R}^n) \to L(H)$ for $X_1, \ldots, X_n$ preserving the almost solvable structure. Let E denote the joint essential spectrum of $X_1, \ldots, X_n$. The representation theorem for $\langle\;\rangle_4$ analogous to that for $\langle\;\rangle_2$ is

<u>Theorem 2.3</u> There exists a continuous linear functional ℓ on $C\Lambda^4(\mathbb{R}^n)$ with compact support such that

$$(f_1, f_2, f_3, f_4)_4 = \mathrm{tr}[\tau(f_1), \tau(f_2), \tau(f_3), \tau(f_4)] = \ell(df_1 \wedge df_2 \wedge df_3 \wedge df_4) .$$

Note if the f_j are just polynomials, then the $\tau(f_j)$ can be taken simply to be $f_j(X_1, \ldots, X_n)$ where the X_i's are inserted in any order. The linear functional ℓ has the property that it vanishes on $E\Lambda^4(\mathbb{R}^n \sim E)$. Thus, ℓ determines an homology class h_ℓ in $H_4(\mathbb{R}^n, E)$.

The index part of this theorem is still under investigation and will be discussed in the next section. Now we mention that h_ℓ determines a third homology class λ_ℓ of E , that is $\lambda_\ell \in H_3(E)$, by Stokes Theorem. Namely, λ_ℓ satisfies

$$(f_1, f_2, f_3, f_4)_4 = \lambda_\ell(\tilde{f}_1 d\tilde{f}_2 \wedge d\tilde{f}_3 \wedge d\tilde{f}_4)$$

where $\tilde{f}_j$ is the restriction of f_j to E .

Here are some examples

(1) For $\mathcal{T}\mathcal{P}_3$ the three sphere S^3 is the set E and h_ℓ for the fundamental term is

$$\frac{1}{\mathrm{vol}(D^4)} \int_{D^4} dz_1 \wedge d\bar{z}_1 \wedge dz_2 \wedge d\bar{z}_2 ,$$

which is the relativized fundamental class of S^3 .

(2) If N is a two-dimensional manifold, then $\mathcal{P}_s N$ has E equal to $S(N)$, the tangent sphere bundle of N and it is a manifold of dimension 3. The representing λ_ℓ for the fundamental form is again in this case the fundamental class of $S(N)$.

§3. Index Theory

The object of this section is to describe our progress toward extending Theorem II.2 to the 2-dimensional case, in other words, we would like to obtain some generalized form of the Atiyah-Singer index theorem for 2-dimensional manifolds. In the 1-dimensional case the index formula was given in terms of h_ℓ an homology class in $H_2(\mathbb{R}^n, E)$. In the 2-dimensional case two homology classes are necessary; one in $H_2(\mathbb{R}^n, E)$, and one in $H_4(\mathbb{R}^n, E)$. The fundamental form will determine the correct homology class h_ℓ in $H_4(\mathbb{R}^n, E)$

(although this isn't entirely proved yet) but there is no simple way of using trace forms to get the lower homology class since in a 2-dimensional algebra tr[A,B] is usually undefined. Section 4 will discuss various ways of obtaining the lower classes.

We begin this section with a discussion of the general index problem and we introduce more machinery than we are able to apply at present. Hopefully, this will help to give a general view of the subject. Recall the notation introduced at the end of §1. Suppose that $X_1,\ldots,X_n$ is a family of algebraic integral operators having C*-algebra $\mathfrak{A}$. The index problem is: write down an explicit formula which gives the index of a Fredholm operator in $M_n \otimes \mathfrak{A}$ in terms of its symbol.

Remark. The algebraic topology relevant to this problem is, briefly, as follows. The index is a homomorphism $\text{ind} : K^1(E) \to \mathbb{Z}$, where E is the joint essential spectrum of the X_i, and K^1 is the usual K-theory functor $[E, G\ell_\infty]$ of homotopy classes of maps from E to the (direct limit of the) general linear groups. Now there is a canonical injection $\text{ch}^* : \text{Hom}(K^1,\mathbb{Z}) \to H_{\text{odd}}(E,Q)$, where ch^* is the (inverse) dual of the usual Chern character. Thus, $\text{ch}^*(\text{ind})$ is a non-homogeneous odd-dimensional rational homology class on E. The problem is to compute this homology class (as explicitly as possible).

An example of an explicit index formula is Theorem II.2. It handles the 1-dimensional case completely. To present more general formulas we define a map $\lambda_k : \Sigma_m(\mathbb{R}^n) \to C\Lambda^{2k}(\mathbb{R}^n,E)$ by

$$\lambda_k(V) = \sum_{i_1,\ldots,i_k} z_{i_1 i_2} \wedge z_{i_2 i_3} \wedge \cdots \wedge z_{i_k i_1}$$

where the z_{ij} are closed two forms given by

$$z_{ij} = \sum_t^n dv_{it} \wedge d\bar{v}_{jt} \ .$$

The v_{ij} are of course the entries of the $m \times m$ matrix valued function V. For example, λ_1 is

$$\lambda_1(V) = \sum_{i,t}^{n} dv_{it} \wedge d\bar{v}_{it}$$

Now let us list the properties of λ_k.

<u>Proposition 3.1</u> If the matrix $s(x)$ is unitary for each point x is an open region $\Omega \subset \mathbb{R}^n$, then $\lambda_k(s)$ is identically zero on Ω.

<u>Idea of Proof</u>: The map λ_k is essentially the pullback of a representative of a 2k-dimensional cohomology class of $M_n(\mathbb{C})$, the $n \times n$ matrices, relative to U_n, the $n \times n$ unitary group. Thus, if $s(x)$ is unitary valued $\lambda_k(s(x))$ is zero.

Suppose that E is a closed subset of $\mathbb{R}^n$. Let $U_m(E)$ denote the group of all matrix functions in $\Sigma_m(\mathbb{R}^n)$ which are unitary in some neighborhood of E. The proposition implies that for any element u of $U_m(E)$, the closed k-form $\lambda_k(u)$ belongs to $C\Lambda^k(\mathbb{R}^n \sim E)$. Thus, the map λ_k induces a map $ch_k : U_n(E) \to H^{2k}(\mathbb{R}^n, E)$ which assigns to u the cohomology class determined by the closed form $\lambda_k(u)$.

The ch_k are, up to normalization, just the homogeneous components of the (relative) Chern character. From algebraic topology, then, we have the following fact.

Suppose that h is an homology class in $H_{2k}(\mathbb{R}^n, E)$. Then for u_1, u_2 in $U_m(\mathbb{R}^n)$

$$h(ch_k(u_1 u_2)) = h(ch_k u_1) + h(ch_k u_2) \quad .$$

This additive property is strongly reminiscent of the additive property of the Fredholm index. In fact, the Fredholm index of a partial isometry in a crypto-integral algebra can be given as the weighted sum of appropriate homology classes applied to the $ch_k(u)$.

Suppose that $X_1,\ldots,X_n$ is a family of algebraic integral operators (dim 2) with joint essential spectrum E . Let $\mathfrak{A}$ denote C*-algebra generated by the family. Let h_4 denote the homology class in $H_4(\mathbb{R}^n,E)$ associated with this family by Theorem 2.3.

Conjecture 3.3 If the operator U in $\mathfrak{A}$ has a symbol which extends from E to a matrix function u in $U_m(\mathbb{R}^n)$ and if $ch_2(u)$ is in $E\Lambda_2(\mathbb{R}^n \sim E)$, then

$$\text{index } U = \frac{1}{12} h_4(ch_4 u) .$$

The authors actually are a good way toward proving this. It is true for TP_3 and $P_s N$ with N a Riemann surface and Proposition 3.2 gives the correct additivity. We can then reduce the problem to one where E is S^3 , $n = 4$, and u is unitary. Hopefully, this case will not prove too difficult.

The condition on $ch_2(u)$ may be mysterious to some people. What is happening is this: from the remark on ch above, we know that there exist homology classes h_2 in $H_2(\mathbb{R}^n,E)$ and h_4 in $H_4(\mathbb{R}^n,E)$ so that

$$\text{(3.1)} \qquad \text{index } U = \frac{1}{12} h_4(ch_4(u)) + h_2(ch_2(u)) .$$

Now if $ch_2(u)$ is exact then the last term vanishes; the conjecture states that h_4 is the homology class h_ℓ determined by the fundamental trace form on $\mathfrak{A}$.

Disclaimer: The normalizations appearing in the index formulas of this section were computed hastily and they should not be taken as binding.

§4. Lower homology classes

The main result of the paper is that the 'trace theory'is

closely related to index theory, when the joint essential spectrum E of the singular integral operators is not bad. This section explores the extent to which the two theories do not give the same information. Briefly the situation is this. As described in §3 the B-D-F work after considerable interpretation says that the index information is determined by a collection of odd homology classes for E and that knowing these classes is equivalent to knowing the index information in the algebra. The trace form in the case we have studied determines only the highest of these homology classes, and most likely the converse is true when E is 'nice' : the highest homology class determines the fundamental trace form. The natural question is: how do you deal with the lower homology classes? This section discusses that topic. The intention is to describe the general situation along with various possibilities and not to present any particular results. The discussion is cast entirely in terms of the one- and two-dimensional cases.

The 'classical' approach to obtaining lower homology class is then 'duality.' Now we describe this. Suppose for the duration of §4 that $X_1,\ldots,X_n$ is a family of algebraic integral operators (dim 2), that $C^\infty\mathfrak{A}$ is a maximal crypto-integral algebra containing them, and that E is the joint essential spectrum of $X_1,\ldots,X_n$. Fix A, B in $C^\infty\mathfrak{A}$; then $\langle\ ,\ , A, B\rangle_4$ is a bilinear form on $C^\infty\mathfrak{A}$. The bilinear form $(\ ,\ )_{A,B}$ which $\langle\ ,\ , A, B\rangle_4$ induces on $C^\infty\mathfrak{A}/\mathfrak{A}_1$ gives a collapsing form on $\partial(R^n)$ which is supported on the joint essential spectrum E of $X_1,\ldots,X_n$. Thus by I, §2, §4 we get an homology class $h_{A,B}$ in $H_2(E,\mathbb{R}^n)$. One question is: can any element of $H_2(E,\mathbb{R}^n)$ be obtained by a suitable choice of A,B ? This question is purely topological, since A,B determine a cohomology class $da \wedge db$ in $H^2(E,\mathbb{R}^n)$ and $h_{A,B}(df \wedge dg) = h_4(df \wedge dg \wedge da \wedge db)$ where h_4 is the homology class canonically arising from the fundamental form $\langle\ \rangle_4$. The answer is yes if E

is a 3-manifold and h_4 is non-trivial (this is called Poincaré duality see Ch. 5 [M]) and no frequently when E is not a manifold. In particular, if E is a manifold, one can express h_2 of formula (3.1) in terms of the fundamental form and in A,B in the algebra. This is the method Atiyah and Singer use in their index theorem and possibly one could do this kind of thing for arbitrary C^∞-extentions on manifolds (not just the pseudo differential operators of order 0). The drawbacks of this method are that it's not clear which A and B to choose; secondly, it only works when E is a manifold and so it cannot be intrinsically operator theoretic. A third problem is that when the fundamental form is identically zero, this entirely breaks down.

The last paragraph dealt with getting homology classes from trace forms. This paragraph deals with getting trace forms from homology classes. The family $X_1,\ldots,X_n$ via the index map (as explained) determines homology classes h_2 and h_4. If the symbols of A and B in $\mathfrak{A}(X_1,\ldots,X_n)$ have extensions a and b in $S(R^n)$ with $da \wedge db$ in $C\Lambda^2(R^n \sim E)$ define the formal trace of $[A,B]$ to be

$$t_2(A,B) = h_2(da \wedge db) .$$

The formal trace is consistent with the actual trace of many commutators. Namely, if X,Y are almost commuting operators in $C^\infty\mathfrak{A}$ with joint essential spectrum M consisting of smooth arcs then $t_2(X,Y) = t_1[X,Y]$. We give the proof. Consider the bilinear form $t_2(\ ,\ )$ on $C^\infty\mathfrak{A}(X,Y)$. By construction it can be expressed in the usual way by an homology class $\tilde{h}_2$ in $H_2(\mathbb{R}^n,M)$ derived from h_2. The definition of h_2 says that index $u = t_2(u,u^*)$ if $u \in C^\infty\mathfrak{A}$ and has a symbol of modulus 1 on E which extends to a function in $C_0^\infty(\mathbb{R}^n)$; thus this will hold for appropriate u in $C^\infty\mathfrak{A}(X,Y)$. The main representation theorem of Part I since M

consists of smooth arcs implies that the fundamental form on $C^\infty\mathfrak{M}(X,Y)$ is completely determined by an element h_2 of $H_2(\mathbb{R}^n,M)$. The fact that $\tilde{h}_2$ and h_2 are the same follows from the comments just made on index. Therefore, $t_2(\ ,\ )$ is equal to the fundamental form and the proof is complete. Note that if X,Y are merely renormalizable and not almost commuting, then $t_2(X,Y) = t(X,Y)$ where $t(\ ,\ )$ is the formal trace from Part I §9. Thus we have seen that the 'index' homology class allows one to define a formal trace. Likewise, one could define formal determinants. It is conceivable to the authors, that these notions of formal trace and determinant will someday have applications.

Now we mention that this process can be reversed to obtain the homology class h_2. Namely, if we are given a renormalizable pair X,Y in $C^\infty\mathfrak{M}$ with 'nice' joint essential spectrum we can in theory compute $t(X,Y)$ and consequently $t_2(X,Y)$. Obviously there are enough renormalizable pairs X,Y in any maximal crypto integral algebra (dim k) to determine h_2 uniquely. It should be emphasized that what is happening here is this. Unless h_4 is trivial one cannot find a compact perturbation of $X_1,\ldots,X_n$ which is almost commuting. However, in non-pathological circumstances, one can find a compact perturbation of any two of the X's which is almost commuting.

§5. C^∞-extensions

The work of Brown, Douglas, and Fillmore deals with extensions of the continuous functions on a set E by the compact operators. Our work could be viewed at a study of extensions of a differentiable manifold E by the trace class operators. Even though we have not pursued this direction, it seems appropriate to mention it in this volume and possibly this absolute setting will come to be regarded as better than the relative setting we have explored.

A maximal crypto-integral algebra (dim k) is one which is properly contained in no other such algebra. Suppose that E is a $2n-1$ dimensional manifold. We say that $(\mathfrak{A}, \phi)$ is an extension of $C^{\infty}(E)$ by the trace class operators provided that $\mathfrak{A}$ is a *-closed maximal crypto-integral algebra (dim n) and ϕ is a *-homomorphism of $\mathfrak{A}$ with the commutator ideal $\mathfrak{A}_1$ in its kernel onto a subalgebra of $C(E)$ which contains $C^{\infty}(E)$. There are many apparent questions to ask. One could try to classify the C^{∞}- extensions up to unitary equivalence in analogy with B-D-F . One can set up a trace theory, since the fundamental form on $\mathfrak{A}$ induces immediately a form $(\ , \ , \ , \ )$ on $C^{\infty}(E)$ which estimates (of the same type used in the other sections) show to be continuous. However, the classification of collapsing forms on manifolds turns out to be more difficult than their classification on $\mathbb{R}^n$. Another reasonable line of work would be to determine the relationship between C^{∞}-extensions and B-D-F extensions. The authors suspect (on the basis of no particular evidence) that each B-D-F extension for E a manifold contains in an appropriate sense a C^{∞}-extension. Possible the C^{∞}-extensions of E depend on its differentiable structures while the B-D-F extensions depend only on its topological structure.

References

[A-S] M. Atiyah and I. M. Singer, Index of elliptic operators I, II, III, Annals of Math. 87 (1968).

[B] I. D. Berg, An extension of the Weyl-Von Neumann Theorem to normal operators, Trans. A.M.S. 160 (1971) 365-371.

[B-D-F] L. Brown, R. G. Douglas, P. Fillmore, Unitary equivalence modulo the compact operators and extensions of C*-algebras, these Notes.

[B-S] C. Berger and B. I. Show, Self-commutators of multi-cyclic hyponormal operators are always trace class, Bull. A.M.S. (See also their paper in these Notes.)

[C_1] K. Clancy, Semi-normal operators with compact self-commutators, Proc. A.M.S. 26 (1970), 447-454.

[C_2] K. Clancy, Examples of non-normal semi-normal operators whose spectra are non-spectral sets, Proc. A.M.S. 24 (1970) 497-800.

[C-F] I. Colojoara, C. Foias, Theory of Generalized Spectral Operators, Gordon and Breach, New York, 1968.

[C-P_1] R. Carey, J. D. Pincus, The structure of intertwining isometries, Indiana Journal 22 (1973) 679-703.

[C-P_2] R. Carey, J. D. Pincus, On an invariant for certain operator algebras (to appear).

[D] R. G. Douglas, Banach algebra techniques in the theory of Toeplitz operators, AMS Region Conf. Series in Math., 15 (1973).

[Dd-S] Deddens and J. Stampfli, On a question of Douglas and Fillmore, Bull. A.M.S. 79 (1973).

[D-S] N. Dunford and J. Schwartz, Linear operators I, Interscience, New York, 1967.

[F] H. Federer, Geometric Measure Theory, Springer, 1969.

[G-K] Gohberg and M. G. Krein, Introduction to the theory of linear non-self adjoint operators, Trans. Math. Monog. A.M.S. 18 (1969).

[H] R. E. Howe, A functional calculus for hyponormal operators, to appear in Indiana Journal.

[K] T. Kato, Smooth operators and commutators, Studia Math. 31 (1968) 535-546.

[M] C.R.F. Maunder, Algebraic Topology, Van Nostrand, New York, 1970.

[M-W] B. McCoy and T. T. Woo, The two-dimensional Ising model, Harvard University Press.

[P_1] J. D. Pincus, Commutators and systems of singular integral equations I, Acta Math. 121 (1968).

[P_2] J. D. Pincus, The determining function method in the treatment of commutator systems, Proc. Intern. Conf. Operator Theory, Tihany (Hungary) (1970).

[P_3] J. D. Pincus, On the trace of commutators in the algebra of operators generated by an algebra with trace class self-commutator (to appear).

[P_4] J. D. Pincus, The spectrum of semi-normal operators, Proc. Nat. Ac. Sci. U.S.A. 68 (1971).

[Put_1] C. R. Putnam, An inequality for the area of hyponormal spectra, Math. Zeitschrift, 116 (1970) 323-330.

[Put_2] C. R. Putnam, Trace normal inequalities of the measure of hyponormal spectral, Indiana Journal 21 (1972).

[Put_3] C. R. Putnam, Commutation properties of Hilbert space operators and related topics, Springer, New York, 1967.

[R] M. Rosenbloom, A spectral theory for self-adjoint singular integral operators, Amer. J. Math. 88 (1966) 314-328.

[Rn] H. Royden, Real Analysis, McMillan, New York, 1964.

[Ru] W. Rudin.

[V] M. Venugopalkrishna, Fredholm operators associated with strongly pseudo-convex domains in C^n, J. Functional Analysis 9 (1972) 349-373.

[X] Xa-Dao-Xeng, On non-normal operators, Chinese Math. 3 (1963) 232-246.

THE DETERMINANT INVARIANT FOR OPERATORS WITH TRACE CLASS SELF COMMUTATORS

Lawrence G. Brown

§1. Introduction

In the previous article, [4], Helton and Howe defined an invariant for operators with trace class self-commutators, by means of traces of commutators, which turns out to give a new interpretation of Pincus' principal function. They also defined another invariant by means of determinants of multiplicative commutators and conjectured a formula for it. Pincus, [7], proved this formula correct in a special case, and we will show that, properly interpreted, it is always correct.

One way of stating our result is that if A and B are invertible operators with $[A,B]$ trace class, then under appropriate hypotheses $\det(ABA^{-1}B^{-1})$ can be computed (in principle) in terms of traces. The hypotheses are:

i) The *-algebra generated by A and B is commutative modulo the trace class; i.e., $[A^*,A], [B^*,B]$ and $[A,B^*]$, as well as $[A,B]$ are trace class.

ii) The joint essential spectrum, X', of A and B is diffeomorphic to a subset of the plane. X' is a compact subset of $\mathbb{C}^2$, and the hypothesis is that there is a compact subset X of the plane and a homeomorphism f of X' onto X such that f is the restriction to X' of a C^∞ function on $\mathbb{C}^2$ and f^{-1} is the restriction to X of a C^∞ function on the plane.

Hypothesis i) does not seem an unduly severe restriction at the present time, but it would be interesting to know what happens if hypothesis ii) is removed. The main case is the case where A and B are unitary, so that X' is a subset of the torus.

§2. Preliminaries

Let H be a separable infinite dimensional Hilbert space, $L(H)$ the algebra of bounded operators on H, K the ideal of compact operators, T the ideal of trace class operators, $A = L(H)/K$ (the Calkin algebra), and $\pi : L(H) \to A$ the projection. If T is an operator whose self-commutator, $[T^*,T] = T^*T-TT^*$, is trace class, let $\mathfrak{A}(T)$ denote the C*-algebra generated by T, I and K, and let $\mathfrak{A}_1$ be a *-subalgebra of $\mathfrak{A}(T)$ containing T and maximal with respect to the property that $R,S \in \mathfrak{A}_1 \Rightarrow [R,S] \in T$. There is a canonical isomorphism between $\pi(\mathfrak{A}(T))$ and $C(X)$, where X is the essential spectrum of T (the spectrum of $\pi(T)$). This gives rise to a "symbol map", $\phi : \mathfrak{A}(T) \to C(X)$. ϕ is surjective and has kernel K. Let Ω_j, $j = 1,2,\ldots,$ be the bounded connected components of the complement of X and let $\tilde{X} = X \cup (\cup_j \Omega_j)$. Also, $f \in C(X)$ is called <u>smooth</u> if f is the restriction to X of a C^∞ function on the plane, and $\mathfrak{A}_2 = \{S \in \mathfrak{A}_1 : \phi(S)$ is smooth$\}$.

Helton and Howe show that for $R,S \in \mathfrak{A}_1$, $\mathrm{tr}[R,S]$ depends only on $\phi(R)$ and $\phi(S)$, and they give a formula for $\mathrm{tr}[R,S]$ for $R,S \in \mathfrak{A}_2$. They construct a real, signed, finite measure m on $\tilde{X}$ such that $\mathrm{tr}[R,S] = \frac{1}{2\pi i} \int J(\tilde{f},\tilde{g})\,dm$, where $\tilde{f}$ and $\tilde{g}$ are C^∞ functions, $\tilde{f}_{|X} = \phi(R)$ and $\tilde{g}_{|X} = \phi(S)$, and $J(\tilde{f},\tilde{g}) = \frac{\partial \tilde{f}}{\partial x}\frac{\partial \tilde{g}}{\partial y} - \frac{\partial \tilde{f}}{\partial y}\frac{\partial \tilde{g}}{\partial x}$. It turns out that m is absolutely continuous with respect to two dimensional Lebesgue measure, which we denote by m_o, and $\frac{dm}{dm_o}(x,y) = -G(y,x)$, where G is Pincus' principal function (see [7], [8] for proof). Helton and Howe also compute m on the Ω_j's, thus giving some information about G

which had not previously been known in so much generality. Before describing this result, we give some more notation. For $S \in \mathfrak{A}(T)$, S is Fredholm if and only if $\phi(S)$ does not vanish anywhere on X. In this case index S depends only on $\phi(S)$ and we denote it by $\gamma(\phi(S))$ (cf. [3]; if necessary to avoid confusion we will write γ_T instead of γ). For f non-vanishing in $C(X)$, $\gamma(f)$ depends only on the homotopy type of f and hence γ gives a homomorphism from $\pi^1(X)$ to $\mathbb{Z}$. ($\pi^1(X)$ is the group of homotopy classes of continuous functions from X to $\mathbb{C} - \{0\}$; also $\pi^1(X) \cong \check{H}^1(X)$.) Now $\pi^1(X)$ is naturally isomorphic to the free abelian group generated by the Ω_j's, and hence γ merely assigns an integer, K_j, to each Ω_j. Explicitly, $K_j = \text{index}(T-\lambda)$ for any $\lambda \in \Omega_j$. Finally, on Ω_j, $m = -K_j m_o$.

For A,B invertible in $\mathfrak{A}_1$, Helton and Howe define $\Delta(A,B) = \det(ABA^{-1}B^{-1})$ and establish the following formal properties:

i) $\Delta(A,B) = \Delta(B,A)^{-1}$

ii) $\Delta(AB,C) = \Delta(A,C)\Delta(B,C)$

iii) $\Delta(A,A) = 1$

iv) $\Delta(A,B)$ is invariant under trace class perturbation of A and B.

For $A,B \in \mathfrak{A}_2$ they conjecture that $\Delta(A,B) = \exp(\frac{1}{2\pi i} \int \frac{J(\tilde{f},\tilde{g})}{\tilde{f}\tilde{g}} dm)$, and this conjecture was proved by Pincus when $\phi(A)$ and $\phi(B)$ have smooth logarithms. It should be noted that $\phi(A)$ and $\phi(B)$ have logarithms if and only if the C^∞ extensions $\tilde{f}$ and $\tilde{g}$ can be taken to be non-vanishing on $\tilde{X}$, and hence in the general case the formula requires some interpretation. Also, we point out that if $E(\ell,z)$ is the determining function of Pincus, then $\det E(\ell,z) = \Delta(\text{Im}T-\ell, \text{Re}T-z)$, and thus the determinant invariant is closely related to one of Pincus' major tools.

§3. The Basic Computations

We start with an invariant intermediate between the trace and determinant invariants. For $A, S \in \mathfrak{A}_1$, A invertible, define $(A,S] = tr(ASA^{-1}-S)$.

Lemma 1.

i) $(A,S]$ depends only on $\phi(A)$ and $\phi(S)$.

ii) $(A,S]$ is linear in S.

iii) $(AB,S] = (A,S] + (B,S]$.

Proof.

i) $ASA^{-1}-S = [AS, A^{-1}]$. Hence by [4], $(A,S]$ depends only on $\phi(AS) = \phi(A)\phi(S)$ and $\phi(A^{-1}) = \phi(A)^{-1}$.

ii) is obvious.

iii) $(AB)S(AB)^{-1}-S = A(BSB^{-1})A^{-1}-BSB^{-1}+BSB^{-1}-S$. Hence $(AB,S] = (A,BSB^{-1}]+(B,S]$. But since $\phi(BSB^{-1}) = \phi(S)$, $(A,BSB^{-1}] = (A,S]$ by i).

Lemma 2. If $A = e^R$, $R \in \mathfrak{A}_1$, then $(A,S] = tr[R,S]$.

Proof. $ASA^{-1} = e^R S e^{-R} = S+[R,S] + \frac{[R,[R,S]]}{2} + \ldots$. Since $[R,S] \in \mathcal{T}$, the series converges in trace norm. For the same reason the higher commutators have trace 0. Therefore $tr(ASA^{-1}-S) = tr[R,S]$.

Lemma 3. For $A, S \in \mathfrak{A}_2$, A invertible, let $\tilde{f}$ be a C^∞ function such that $\tilde{f}_{|X} = \phi(A)$ and let $\tilde{g}$ be a C^∞ function such that $\tilde{g}_{|X} = \phi(S)$ and $\tilde{g}$ vanishes in a neighborhood of $K_{\tilde{f}} = \{x \in \tilde{X} : \tilde{f}(x) = 0\}$. Then $(A,S] = \frac{1}{2\pi i} \int \frac{J(\tilde{f},\tilde{g})}{\tilde{f}} \, dm$.

Remarks. 1. $K_{\tilde{f}}$ is a compact set disjoint from X. Hence $\tilde{g}$ always exists.

2. Another version of the formula for $(A,S]$ will be given in §5.

Proof of Lemma 3. Let $\tilde{g}$ vanish in a neighborhood U of $K_{\tilde{f}}$, and let K' be a compact neighborhood of $K_{\tilde{f}}$ contained in $U-X$. Let $\tilde{h}$ be a C^{∞} function such that $\tilde{h}$ agrees with $\frac{1}{\tilde{f}}$ in $V-K'$ for some neighborhood V of $\tilde{X}$. Then $\tilde{h}_{|X} = \phi(A^{-1})$, and hence by [4] $(A,S] = tr[AS,A^{-1}] = \frac{1}{2\pi i}\int J(\tilde{f}\tilde{g},\tilde{h})dm$. On $V-K'$, $J(\tilde{f}\tilde{g},\tilde{h}) =$ $= J(\tilde{f}\tilde{g},\tilde{f}^{-1}) = \tilde{g}J(\tilde{f},\tilde{f}^{-1}) + \tilde{f}J(\tilde{g},\tilde{f}^{-1}) = 0 - \tilde{f}\cdot\tilde{f}^{-2}J(\tilde{g},\tilde{f}) = \frac{J(\tilde{f},\tilde{g})}{\tilde{f}}$. On U, $J(\tilde{f}\tilde{g},\tilde{h}) = 0 = \frac{J(\tilde{f},\tilde{g})}{\tilde{f}}$. Hence $\frac{1}{2\pi i}\int J(\tilde{f}\tilde{g},\tilde{h})dm =$ $\frac{1}{2\pi i}\int \frac{J(\tilde{f},\tilde{g})}{\tilde{f}}\,dm$.

Lemma 4. If $A,R \in \mathfrak{A}_1$, A invertible, then $\Delta(A,e^R) = \exp((A,R])$.

Proof. $Ae^RA^{-1}e^{-R} = \exp(ARA^{-1})\cdot\exp(-R)$. If $||R||$ is sufficiently small, we can apply the Campbell-Baker-Hausdorff formula to obtain $Ae^RA^{-1}e^{-R} = \exp(ARA^{-1}-R - \frac{1}{2}[ARA^{-1},R] + \ldots)$. Since $[ARA^{-1},R] \in \mathcal{T}$, the series converges in trace norm (if $||R||$ is sufficiently small, for fixed A). For the same reason, the higher commutators have trace 0. Also, since $ARA^{-1}-R \in \mathcal{T}$, $tr[ARA^{-1},R] = tr[ARA^{-1}-R,R]=0$. Hence $\det(Ae^RA^{-1}e^{-R}) = \exp(tr(ARA^{-1}-R)) = \exp(A,R]$. If $||R||$ is not small, we use the fact that $\Delta(A,e^R) = \Delta(A,(e^{R/n})^n) = \Delta(A,e^{R/n})^n$ and $\exp((A,R]) = \exp(n(A,R/n]) = (\exp(A,R/n])^n$. Hence by choosing n large enough, we can deduce the result from the case proved above.

Corollary 4.1. $\Delta(A,B)$ depends only on $\phi(A)$ and $\phi(B)$.

Proof. By the formal properties of Δ, we need only prove that $\Delta(A,B) = 1$ if $A = 1+K$ for some $K \in \mathfrak{A}_1 \cap \mathcal{K}$. Further, by replacing A with a finite rank perturbation, we may assume $||K|| < 1$. Then $A = e^L$ for some $L \in \mathfrak{A}_1 \cap \mathcal{K}$. Hence $\Delta(A,B) =$ $= \Delta(B,A)^{-1} = \exp(-(B,L]) = 1$, by i) of Lemma 1.

Remark. At this point we could prove that the conjectured formula for $\Delta(A,B)$, properly interpreted, is correct if either $\phi(A)$ or $\phi(B)$ has a logarithm. This would be an improvement over Pincus' result, where both are assumed to have logarithms. We are not doing this now because we want to use a more sophisticated interpretation of the conjectured formula.

We are now justified in defining a form δ by $\delta(\phi(A),\phi(B)) = \Delta(A,B)$, for A,B invertible elements of $\mathfrak{A}_2$. It should be noted that the domain of δ is the set of all pairs (f,g) of smooth non-vanishing functions on X such that $\gamma(f) = \gamma(g) = 0$. δ satisfies properties i), ii), iii) of §2.

§4. Direct Sums

Suppose $T_1 \in L(H_1)$ and $T_2 \in L(H_2)$ have trace class self-commutators and $T = T_1 \oplus T_2 \in L(H_1 \oplus H_2)$. Choose the algebra $\mathfrak{A}_1$ for T so that $\mathfrak{A}_1 \supset \{S_1 \oplus S_2 : S_1 \in \mathfrak{A}_1(T_1),\ S_2 \in \mathfrak{A}_1(T_2)$, and $\phi_1(S_1) = \phi_2(S_2)$ on $X_1 \cap X_2\}$. If m_1, m_2 and m are the measures defining the trace invariants for T_1, T_2 and T, then clearly $m = m_1 + m_2$. Also, for non-vanishing $f \in C(X)$, $\gamma_T(f) = \gamma_{T_1}(f|_{X_1}) + \gamma_{T_2}(f|_{X_2})$; and for smooth non-vanishing functions $f,g \in C(X)$, $\delta(f,g) = \delta_1(f|_{X_1}, g|_{X_1}) \cdot \delta_2(f|_{X_2}, g|_{X_2})$ whenever the right-hand side is defined.

Further, choose an orthonormal basis of H and for $A \in L(H)$ define $A^t \in L(H)$ as the operator whose matrix with respect to that basis is the transpose of the matrix of A. The transpose operation so defined depends on the choice of the basis, but changing the basis would not change the unitary equivalence class of A^t. Now if T has trace class self-commutator and essential spectrum X and if $T' = T^t$, then T' also has trace class self-commutator and essential spectrum X. Clearly $m' = -m$, $\gamma' = -\gamma$, and $\delta' = \delta^{-1}$. (Note that δ and δ' have the same domain.)

Finally, if Ω is a bounded open set in $\mathbb{C}$, Berger and Shaw

[1], [2] have shown that multiplication by z on $A^2(\Omega)$ gives a hyponormal operator T_Ω with trace class self-commutator and $\mathrm{tr}[T_\Omega^*, T_\Omega] = \frac{1}{\pi} m_o(\Omega)$. The essential spectrum of T_Ω is contained in $\partial\Omega$, the boundary of Ω , $\mathrm{index}(T_\Omega - \lambda) = -1$ for $\lambda \in \Omega$, and $\mathrm{index}\,(T_\Omega - \lambda) = 0$ for λ not in $\mathrm{Cl}(\Omega)$. Helton and Howe [4] show that for a general T with trace class self-commutator, $||m|| \leq \pi||[T^*, T]||_1$. In particular, $\Sigma_j |K_j| m_o(\Omega_j) \leq \pi||[T^*, T]||_1$, and if equality holds m vanishes on X . Of course, equality does hold for T_Ω . More generally, let X be an arbitrary compact subset of $\mathbb{C}$, let the Ω_j be as usual, and let K_j be any given sequence of integers such that $\Sigma_j |K_j| m_o(\Omega_j) < \infty$. Let T_j be the direct sum of T_{Ω_j} with itself $|K_j|$ times if $K_j \leq 0$, and the direct sum of $T^t_{\Omega_j}$ with itself K_j times if $K_j \geq 0$. Let $T = \Sigma_j \oplus T_j$. Then T has trace class self-commutator and essential spectrum a subset of X . If we take the direct sum of T with a normal operator having essential spectrum X , we obtain an operator T' with the given K_j's and with measure m' vanishing on X .

Combining the three above remarks, we see that the general problem can be reduced to the case where m either vanishes on X or is concentrated on X . In fact, if T is given and T' is the operator constructed above (with the same K_j's as T), then consider $T \oplus T'^t$ and T' . The determinant invariants that arise (δ, δ', etc.) all have the same domain. Other reductions can also be done by using the ideas of this section.

§5. Topological Considerations and the Definition of d

Let f be a smooth non-vanishing function on X . To evaluate the conjectured formula for δ , we need to extend f to a C^∞ function $\tilde{f}$ on a neighborhood of $\tilde{X}$. $\tilde{f}$ might vanish somewhere and this is the problem that necessitates interpreting the formula. We now consider this in detail.

First, let $\tilde{f}'$ be any C^{∞} extension. Then there is some neighborhood U of X in which $\tilde{f}'$ does not vanish. U already includes all but finitely many Ω_j's. Therefore we need only try to modify $\tilde{f}'$ in finitely many Ω_j's in order to eliminate the zeroes. (Note that it is not necessary to consider $\tilde{f}'$ outside of $U \cup \tilde{X}$.) Now for a fixed j, let Ω be a domain with smooth regular boundary such that $\Omega \supset \Omega_j - U$ and $Cl(\Omega) \subset \Omega_j$. In particular $\partial\Omega \subset U$. The restriction of $\tilde{f}'$ to $\partial\Omega$ has some winding number, N_j, about 0. If $N_j = 0$, then we can modify $\tilde{f}'$ in Ω so that the new function is still C^{∞} and does not vanish in Ω_j. If $N_j \neq 0$, then we cannot do this, even if the new function were only required to be continuous. Instead we choose a point $z_j \in \Omega$ and a small closed disk D_j centered at z_j. Let f_j be any non-vanishing C^{∞} function on $D_j - \{z_j\}$ such that the winding number of $f_{j|\partial D_j}$ about 0 is N_j. Then we can modify $\tilde{f}'$ in Ω so that the new function is C^{∞} and non-vanishing in $\Omega_j - \{z_j\}$ and agrees with f_j on $D_j - \{z_j\}$.

If we perform these modifications in each Ω_j in which $\tilde{f}'$ has zeroes, we obtain a non-vanishing C^{∞} function $\tilde{f}$, with finitely many isolated singularities in $\tilde{X}-X$, such that $\tilde{f}_{|X} = f$. ($\tilde{f}$ is defined on a neighborhood of $\tilde{X}$ except for the singularities.) In general, if we have such an $\tilde{f}$ and if z_0 is one of the singularities, then we say that the <u>order</u> of z_0, denoted $o(z_0)$, is the winding number about 0 of the restriction of $\tilde{f}$ to a small simple closed curve about z_0. Also, if $z_0 \in \Omega_j$, then we define $K(z_0) = K_j$. Note that we can prescribe the nature of the singularities of the extended function $\tilde{f}$, provided only that we allow at least one type of singularity of each order; and if we choose a point z_j in each Ω_j, we can stipulate that the singularities of $\tilde{f}$ occur only at some of the z_j's. For functions $\tilde{f}$ of this sort, we will call the singularities the <u>zeroes</u> of $\tilde{f}$, regardless of the actual nature of the singularities. Note that from our present point of view any actual zeroes of $\tilde{f}$ in $\tilde{X}$ must be regarded as singularities (since $\tilde{f}$ is required to be non-zero where defined; zeroes outside $\tilde{X}$ will be ignored, since they can be eliminated by changing to a smaller neighborhood of $\tilde{X}$).

Now with $\tilde{f}$ as above, having zeroes at $z_1,\ldots,z_\ell$, define $N_j = \sum_{z_i \in \Omega_j} o(z_i)$. All but finitely many N_j's are 0. The N_j's are topological invariants of $f = \tilde{f}_{|X}$. In fact, we have already pointed out that $\pi^1(X)$ is naturally isomorphic to the free abelian group generated by the Ω_j's. Hence the image of f in $\pi^1(X)$ gives an integer for each Ω_j, all but finitely many of the integers being 0. These integers are precisely the N_j's. (The reader might want to consult [5], section 6-17, in connection with the remarks made so far.) In particular $\gamma(f) = \Sigma_j N_j K_j = \sum_{i=1}^{\ell} o(z_i)K(z_i)$. Also, f has a logarithm if and only if all N_j's $= 0$. In fact if all N_j's $= 0$, $\tilde{f}$ can be taken to have no zeroes. Now since $\mathbb{C}-\tilde{X}$ is connected, $\tilde{X}$ has arbitrarily small simply connected neighborhoods. If $\tilde{f}$ has no zeroes, then the restriction of $\tilde{f}$ to one of these neighborhoods has a C^∞ logarithm. Conversely, if f has a (continuous) logarithm, clearly the image of f in $\pi^1(X)$ is trivial, and hence all N_j's $= 0$.

We do need to put some restriction on the types of zeroes $\tilde{f}$ can have. If $\tilde{f}$ is as above with zeroes at $z_1,\ldots,z_\ell$, we say $\tilde{f}$ is <u>admissible</u> if for each i there are a $p_i < 2$ and $M_i > 0$ such that

$$\frac{\sqrt{\left|\frac{\partial \tilde{f}}{\partial x}\right|^2 + \left|\frac{\partial \tilde{f}}{\partial y}\right|^2}}{|\tilde{f}(z)|} \leq \frac{M_i}{|z-z_i|^{p_i}}$$

for z sufficiently near z_i .

<u>Remark</u>. Among the types of zeroes permitted are $g(z)(z-z_0)^n$ and $g(z)(\bar{z}-\bar{z}_0)^n$, where g is a C^∞ function not vanishing at z_0 . These zeroes have order n and $-n$, respectively. If we use these for $n > 0$, we get "zeroes" of all orders that are literally zeroes, where the function $\tilde{f}$ will be C^∞ . On the other hand, the zeroes of the form $g(z)(z-z_0)^n$, n positive or negative, also include all orders. These are more convenient because they can occur as the zeroes of holomorphic functions. Holomorphic functions are handy to use because the Jacobian of any two is 0 . We have made the definition of admissibility fairly permissive because we feel that if the determinant invariant has any applications, the machinery should allow considerable flexibility.

<u>Lemma 5</u>. If Ω is a bounded open set with regular smooth boundary, $\tilde{f}$ is admissible on a neighborhood of $Cl(\Omega)$ with no zeroes on $\partial\Omega$, and $\tilde{g}$ is C^∞ on a neighborhood of $Cl(\Omega)$, then

$$-\frac{1}{2\pi i}\int_{\partial\Omega} \tilde{g}\,\frac{d\tilde{f}}{\tilde{f}} = \frac{1}{2\pi i}\int_{\Omega} \frac{J(\tilde{f},\tilde{g})}{\tilde{f}}\,dm_o - \sum_{i=1}^{\ell} o(z_i)\tilde{g}(z_i) ,$$

where $z_1, z_2, \ldots, z_\ell$ are the zeroes of $\tilde{f}$ in Ω .

<u>Proof</u>. Note that $\int \frac{J(\tilde{f},\tilde{g})}{\tilde{f}}\,dm_o$ is absolutely convergent. For small $\varepsilon > 0$, let $D_i(\varepsilon)$ be the closed disk of radius ε and center z_i . Let $\Omega_\varepsilon = \Omega - \bigcup_{i=1}^{\ell} D_i(\varepsilon)$. Then

$$\frac{1}{2\pi i}\int_{\Omega} \frac{J(\tilde{f},\tilde{g})}{\tilde{f}}\,dm_o = \lim_{\varepsilon\to 0} \frac{1}{2\pi i}\int_{\Omega_\varepsilon} \frac{d\tilde{f}\wedge d\tilde{g}}{\tilde{f}}$$

$$= \lim_{\varepsilon\to 0}\left(-\frac{1}{2\pi i}\right)\int_{\Omega_\varepsilon} d\left(\tilde{g}\,\frac{d\tilde{f}}{\tilde{f}}\right) = \lim_{\varepsilon\to 0}\left(-\frac{1}{2\pi i}\right)\int_{\partial\Omega_\varepsilon} \tilde{g}\,\frac{d\tilde{f}}{\tilde{f}}$$

$$= -\frac{1}{2\pi i}\int_{\partial\Omega} \tilde{g}\,\frac{d\tilde{f}}{\tilde{f}} + \sum_{i=1}^{\ell} \lim_{\varepsilon\to 0} \frac{1}{2\pi i}\int_{\partial D_i(\varepsilon)} \tilde{g}\,\frac{d\tilde{f}}{\tilde{f}}$$

$$= -\frac{1}{2\pi i}\int_{\partial\Omega}\tilde{g}\,\frac{d\tilde{f}}{\tilde{f}} + \sum_{i=1}^{\ell} o(z_i)\tilde{g}(z_i) \ .$$

Here we have used Stokes' theorem, and the inequalities in the definition of admissibility are adequate for evaluating

$$\lim_{\varepsilon\to 0}\frac{1}{2\pi i}\int_{\partial D_i(\varepsilon)}\tilde{g}\,\frac{d\tilde{f}}{\tilde{f}} \ .$$

Corollary 5.1. If $A,S \in \mathfrak{A}_2$, A invertible, $\tilde{f}$ is an admissible function such that $\tilde{f}_{|X} = \phi(A)$, and $\tilde{g}$ is any C^∞ function such that $\tilde{g}_{|X} = \phi(S)$, then $(A,S] = \frac{1}{2\pi i}\int \frac{J(\tilde{f},\tilde{g})}{\tilde{f}}\,dm + \sum_{i=1}^{\ell} o(z_i)K(z_i)\tilde{g}(z_i)$, where $z_1, z_2, \ldots$ are the zeroes of $\tilde{f}$.

Proof. Choose small $\varepsilon > 0$ and let $\Omega = \operatorname{int}\bigcup_{i=1}^{\ell} D_i(\varepsilon)$. Let $\tilde{f}'$ be an admissible function, C^∞ on all of $\tilde{X}$, which has a zero at each z_i of the same order as $\tilde{f}$ and no other zeroes, and which agrees with $\tilde{f}$ outside Ω. Let $\tilde{g}'$ be a C^∞ function which vanishes in a neighborhood of each z_i and agrees with $\tilde{g}$ outside Ω. Then, by Lemma 3,

$$(A,S] = \frac{1}{2\pi i}\int \frac{J(\tilde{f}',\tilde{g}')}{\tilde{f}'}\,dm = \frac{1}{2\pi i}\int_{\tilde{X}-\Omega}\frac{J(\tilde{f}',\tilde{g}')}{\tilde{f}'}\,dm$$

$$+ \frac{1}{2\pi i}\int_{\Omega}\frac{J(\tilde{f}',\tilde{g}')}{\tilde{f}'}\,dm$$

$$= \frac{1}{2\pi i}\int_{\tilde{X}-\Omega}\frac{J(\tilde{f},\tilde{g})}{\tilde{f}} - \sum_{i=1}^{\ell}\frac{K(z_i)}{2\pi i}\int_{D_i(\varepsilon)}\frac{J(\tilde{f}',\tilde{g}')}{\tilde{f}'}\,dm_o$$

$$= \frac{1}{2\pi i}\int_{\tilde{X}-\Omega}\frac{J(\tilde{f},\tilde{g})}{\tilde{f}}\,dm + \sum_{i=1}^{\ell} K(z_i)\,\frac{1}{2\pi i}\int_{\partial D_i(\varepsilon)}\tilde{g}'\,\frac{d\tilde{f}'}{\tilde{f}'}$$

$$= \frac{1}{2\pi i}\int_{\tilde{X}-\Omega}\frac{J(\tilde{f},\tilde{g})}{\tilde{f}}\,dm + \sum_{i=1}^{\ell}\frac{K(z_i)}{2\pi i}\int_{\partial D_i(\varepsilon)}\tilde{g}\,\frac{d\tilde{f}}{\tilde{f}}$$

$$= \frac{1}{2\pi i}\int_{\tilde{X}-\Omega}\frac{J(\tilde{f},\tilde{g})}{\tilde{f}}\,dm - \sum_{i=1}^{\ell}\frac{K(z_i)}{2\pi i}\int_{D_i(\varepsilon)}\frac{J(\tilde{f},\tilde{g})}{\tilde{f}}\,dm_o$$

$$+ \sum_{i=1}^{\ell} K(z_i)o(z_i)\tilde{g}(z_i)$$

$$= \frac{1}{2\pi i} \int \frac{J(\tilde{f},\tilde{g})}{\tilde{f}} dm + \sum_{i=1}^{\ell} o(z_i)K(z_i)\tilde{g}(z_i) .$$

Now for $\tilde{f}$ and $\tilde{g}$ admissible with zeroes at $z_1,\dots,z_\ell$ and $w_1,\dots,w_m$ respectively and $z_i \neq w_k$ for all i,k, define

$$D(\tilde{f},\tilde{g}) = \exp\left(\frac{1}{2\pi i} \int \frac{J(\tilde{f},\tilde{g})}{\tilde{f}\tilde{g}} dm\right) \cdot \prod_{i=1}^{\ell} \tilde{g}(z_i)^{o(z_i)K(z_i)} \cdot \prod_{k=1}^{m} \tilde{f}(w_k)^{-o(w_k)K(w_k)}$$

It is clear that $D(\tilde{g},\tilde{f}) = D(\tilde{f},\tilde{g})^{-1}$, and $D(\tilde{f}_1\tilde{f}_2,\tilde{g}) = D(\tilde{f}_1,\tilde{g})\cdot D(\tilde{f}_2,\tilde{g})$ whenever the right-hand side is defined. The integral is absolutely convergent.

Corollary 5.2. If f and g are non-vanishing smooth functions on X such that $\gamma(f) = \gamma(g) = 0$, $\tilde{f}$ and $\tilde{g}$ are admissible functions such that $\tilde{f}_{|X} = f$ and $\tilde{g}_{|X} = g$, and if either $\tilde{f}$ or $\tilde{g}$ has no zeroes, then $\delta(f,g) = D(\tilde{f},\tilde{g})$.

Proof. We may assume that $\tilde{g}$ has no zeroes, so that $\tilde{g} = \exp(\tilde{h})$ for some C^∞ $\tilde{h}$. Let A and S be elements of $\mathfrak{A}_2$, A invertible, such that $\phi(A) = f$ and $\phi(S) = \tilde{h}_{|X}$. Then

$$\delta(f,g) = \Delta(A,e^S) = \exp((A,S])$$

$$= \exp\left(\frac{1}{2\pi i} \int \frac{J(\tilde{f},\tilde{g})}{\tilde{f}\tilde{g}} dm + \sum_{i=1}^{\ell} o(z_i)K(z_i)\tilde{h}(z_i)\right)$$

$$= \exp\left(\frac{1}{2\pi i} \int \frac{J(\tilde{f},\tilde{g})}{\tilde{f}\tilde{g}} dm\right) \prod_{i=1}^{\ell} \exp(\tilde{h}(z_i))^{o(z_i)K(z_i)} = D(\tilde{f},\tilde{g}) .$$

Remark. Given f and g, $\tilde{f}$ and $\tilde{g}$ exist as in Corollary 5.2 if and only if either f or g has a logarithm.

Lemma 6. $D(\tilde{f},\tilde{g})$ depends only on $\tilde{f}_{|X}$ and $\tilde{g}_{|X}$.

Proof. It is sufficient to show that if $\tilde{f}_{1|X} = \tilde{f}_{2|X}$ and $D(\tilde{f}_1,\tilde{g})$ and $D(\tilde{f}_2,\tilde{g})$ are both defined, then $D(\tilde{f}_1,\tilde{g}) = D(\tilde{f}_2,\tilde{g})$. To prove this we choose $\tilde{f}'$ such that $D(\tilde{f}',\tilde{g})$ is defined and $\tilde{f}'_{|X} = \tilde{f}^{-1}_{1|X}$. Then it is sufficient to show $D(\tilde{f}_1\tilde{f}',\tilde{g}) = D(\tilde{f}_2,\tilde{f}',\tilde{g}) = 1$. In other words, we are reduced to showing $D(\tilde{f},\tilde{g}) = 1$ if $\tilde{f}_{|X} = 1$.

We first reduce to the case where $\tilde{f}$ has no zeroes. Let D be a closed disk contained in $\tilde{X}-X$ such that $\tilde{g}$ has no zeroes in D and $\tilde{f}$ has no zeroes on ∂D . It should be clear that if we modify $\tilde{f}$ however we like in int D , so that it remains admissible, we do not change $D(\tilde{f},\tilde{g})$. To see this, choose a logarithm $\tilde{h}$ of $\tilde{g}_{|D}$ and use Lemma 5. Now fix a j such that $\tilde{f}$ has a zero in Ω_j and recall that $N_j(f) = 0$. Choose a sequence $D_1,\ldots,D_k$ of disks in Ω_j satisfying the above hypothesis such that $(\text{int } D_i) \cap (\text{int } D_{i+1}) \neq \emptyset$ and every zero of $\tilde{f}$ in Ω_j is in one of the D_i's . First modify $\tilde{f}$ on int D_1 , so that all zeroes of the new function in int D_1 lie in int $D_1 \cap$ int D_2 . Then modify $\tilde{f}$ on $D_2,D_3,\ldots,D_{k-1}$, to obtain a new function all of whose zeroes in Ω_j are in int D_k . Finally, use the fact that $N_j = 0$ to modify $\tilde{f}$ in int D_k to eliminate all zeroes in Ω_j . By repeating this process for finitely many j's, we achieve the desired reduction.

Now clearly $\gamma(\tilde{f}_{|X}) = 0$. If also $\gamma(\tilde{g}_{|X}) = 0$, then the conclusion follows from Corollary 5.2.

In the general case we use a trick. Let D be a closed disk disjoint from $\tilde{X}$, with center a and radius r . If U_o is the unilateral shift, consider $T' = T \oplus (a + rU_o)$. Note that $\tilde{X}' = \tilde{X} \cup D$. Define an admissible function $\tilde{f}'$ by letting $\tilde{f}' = \tilde{f}$ in a neighborhood of $\tilde{X}$ and $\tilde{f}' = 1$ in a neighborhood of D . Define $\tilde{g}'$ so that $\tilde{g}' = \tilde{g}$ in a neighborhood of $\tilde{X}$ and $\gamma(\tilde{g}'_{|X'}) = 0$. Then by the above $D'(\tilde{f}',\tilde{g}') = 1$. But obviously $D'(\tilde{f}',\tilde{g}') = D(\tilde{f},\tilde{g})$.

We can now define d by $d(\tilde{f}_{|X},\tilde{g}_{|X}) = D(\tilde{f},\tilde{g})$. The domain of d is the set of all pairs (f,g) of non-vanishing smooth functions on X . Sometimes we will write $d(f',g')$ instead of $d(f'|_X, g'|_X)$. In such cases $D(f',g')$ will not necessarily be defined. Clearly d satisfies formal properties i) and ii) of §2. We can now state our theorem whose proof will be given in §6.

Theorem. d agrees with δ on the domain of δ .

By Corollary 5.2 we know the theorem is true for pairs (f,g) such that either f or g has a logarithm. The reason we have allowed d to have a bigger domain than δ is that this facilitates computations. For example, if $T = T_1 \oplus T_2$, as in §4, then $d(f,g) = d_1(f|_{X_1}, g|_{X_1}) \cdot d_2(f|_{X_2}, g|_{X_2})$ whenever the left-hand side is defined. Recall that in the analogous equation for δ , the left side could be defined at places where the right side is not defined. We know of no direct (not involving direct sums) operator theoretic interpretation of d outside the domain of δ . If we had such an interpretation, the proof of the theorem would probably be considerably easier.

Lemma 7. $d(f,f) = (-1)^{\gamma(f)}$.

Proof. There is a rational function r , whose poles and zeroes are in $\tilde{X}-X$, such that r has prescribed "zeroes" and orders. Since $f = (fr^{-1})r$, we can write $f = f_1 \cdot f_2 \cdots f_n$, where each f_i either has a logarithm or is of the form $(z-a)^m$ for some $a \in \tilde{X}-X$. Since $d(f_i,f_j) = d(f_j,f_i)^{-1}$, $d(f,f) = \prod_{i=1}^{n} d(f_i,f_i)$. If f_i has a logarithm, $d(f_i,f_i) = \delta(f_i,f_i) = 1$. Thus we need only prove: $d(z-a,z-a) = (-1)^{K(a)}$. Let D be a closed disk of radius r and center a such that $D \cap X = \emptyset$, let $0 < r_1 < r$, and let D_1 be the closed disk of radius r_1 and center a . Choose a sequence $a_\ell \neq a$ in $\operatorname{int} D_1$ converging to a . For each ℓ , choose a function $\tilde{g}_\ell$, C^∞ on $\mathbb{C} - \{a_\ell\}$, such that $\tilde{g}_\ell$ agrees with $z-a$

outside D , $\tilde{g}_\ell = (z-a_\ell)$ in D_1 , and $\tilde{g}_\ell$ is non-vanishing in $D-D_1$. Also, since $(z-a_\ell)_{|\partial D_1}$ is close in the C^∞ topology to $(z-a)_{|\partial D_1}$, for large ℓ , we may choose $\tilde{g}_\ell$ so that $\tilde{g}_{\ell|D-D_1}$ converges to $(z-a)_{|D-D_1}$ in the C^1 topology. Then $J(z-a,\tilde{g}_\ell) = 0$ outside $D-D_1$ and converges uniformly to 0 on $D-D_1$. Thus

$$d(z-a,z-a) = D(z-a,\tilde{g}_\ell) = \lim_{\ell\to\infty} D(z-a,\tilde{g}_\ell) = \lim_{\ell\to\infty} \tilde{g}_\ell(a)^{K(a)} \cdot (a_\ell-a)^{-K(a_\ell)}$$

$$= \lim_{\ell\to\infty} (a-a_\ell)^{K(a)}(a_\ell-a)^{-K(a)} = (-1)^{K(a)} .$$

Corollary 7.1. $d(f,f) = 1$ if $\delta(f,f)$ is defined.

Finally, we point out that it can easily be proved that $d(f,g) = \exp(\frac{1}{2\pi i}\int \frac{J(\tilde{f},\tilde{g})}{\tilde{f}\tilde{g}}\, dm)$, for any C^∞ functions $\tilde{f}$ and $\tilde{g}$ (not necessarily admissible) such that $\tilde{f}_{|X} = f$, $\tilde{g}_{|X} = g$, $\tilde{f} = 1$ in a neighborhood of $K_{\tilde{g}}$ and $\tilde{g} = 1$ in a neighborhood of $K_{\tilde{f}}$. This gives us an interpretation of $d(f,g)$ similar to the formula for $(A,S]$ in Lemma 3.

§6. Proof of the Theorem

Lemma 8. If m is concentrated on X , $d = \delta$.

Proof. In this case γ and the K_j's are 0 , so that d and δ have the same domain. Thus (cf. the proof of Lemma 7) it is sufficient to prove $\delta(z-a,z-b) = d(z-a,z-b) = 1$, for $a \neq b \in \tilde{X}-X$. Now $T-a$ and $T-b$ are both Fredholm operators of index 0 . Hence there is a finite rank perturbation T' of T such that $T'-a$ and $T'-b$ are both invertible. Thus $\delta(z-a,z-b) = \Delta(T'-a,T'-b) = 1$ since $T'-a$ and $T'-b$ commute.

Corollary 8.1. $m = 0$ implies $\delta = 1$.

Corollary 8.2. δ depends only on m and X .

Proof. Let T_1 and T_2 have trace class self-commutators, the same essential spectrum X, and $m_1 = m_2$. Then δ_1 and δ_2 have the same domain. Consider $T = T_1 \oplus T_2^t$. Since $m = m_1 - m_2 = 0$, $\delta = 1$. But δ agrees with $\delta_1 \cdot \delta_2^{-1}$ on the domain of δ_1. Hence $\delta_1 = \delta_2$.

Remark. It would be nice to say simply "δ depends only on m", but this does not make sense because the domain of δ depends on X. To the extent that it does make sense, it is true.

Proof of the Theorem. It is sufficient to prove $\delta(f,g) = d(f,g)$ for f and g rational functions with poles and zeroes in $\tilde{X}-X$. Let $z_1,\ldots,z_\ell$ be the poles and zeroes of f and g. Choose open disks D_i centered at z_i such that the $Cl(D_i)$'s are disjoint from each other and from X. Let $X' = \tilde{X} - \sum_{i=1}^{\ell} D_i$, let N be a normal operator with essential spectrum X', and consider $T' = T \oplus N$. Since $\delta'(f,g) = \delta(f,g)$ and $d'(f,g) = d(f,g)$, we may work with T' instead of T. In other words we are reduced to the case where there are only finitely many Ω_j's, each a disk, having disjoint closures.

Now let D_o be a closed disk disjoint from $\tilde{X}$ with center a_o and radius r_o, and replace T with $T \oplus (a_o + r_o U_o)$, as in the proof of Lemma 6. The advantage of this is that $K_o = -1$. Thus if a_j is the center of Ω_j and $f_j = (z-a_j)(z-a_o)^{K_j}$, then $\gamma(f_j) = 0$ and it is sufficient to prove (for the new T) $\delta(f_i,f_j) = d(f_i,f_j)$.

Thus we need only consider three Ω_j's, and we are reduced to the case where there are only three Ω_j's, which are D', D'', D''', where D', D'' and D''' are open disks whose closures are disjoint and $K' = -1$. Now in §4 we explained how to reduce to the case where m is either concentrated on X or vanishes on X, and in Lemma 8 we did the case where m is concentrated on X. Thus we may assume m vanishes on X. Also, by Corollary 8.2, we need consider only a single "model operator" T which gives rise to m. Let the centers and radii of the disks be a', a'', and a''' and r', r'', r''' respectively. Let U' be a unilateral shift on $\mathcal{H}'$ and U'' (respectively U''') a forward or backward unilateral shift, according as K'' (respectively K''') is negative or positive, on $\mathcal{H}''$ (respectively $\mathcal{H}'''$). Let $T' = a'+r'U'$, $T'' = a''+r''U''$ and $T''' = a'''+ r'''U'''$. We may assume that T is the direct sum of T', $|K''|$ copies of T'', $|K'''|$ copies of T''', and a normal operator. The normal summand is clearly irrelevant. Hence we may ignore it and reduce to the case where $\tilde{X}$ is the union of the three closed disks. Let $f'' = (\frac{z-a'}{r'})^{|K''|}$ on D', $(\frac{z-a''}{r''})^{\operatorname{sgn}K''}$ on D'', and 1 on D'''; and let $f''' = (\frac{z-a'}{r'})^{|K'''|}$ on D', 1 on D'' and $(\frac{z-a'''}{r'''})^{\operatorname{sgn}K'''}$ on D'''. It is sufficient to show $\delta(f'',f''') = d(f'',f''')$. Note that $d(f'',f''') = (-1)^{K''K'''}$. Let $A_1 = U'\oplus V''\oplus I\oplus\ldots\oplus I$, $A_2 = U'\oplus I\oplus V''\oplus I\oplus\ldots\oplus I,\ldots,A_{|K''|}=U'\oplus I\oplus\ldots\oplus V''\oplus I\oplus\ldots\oplus I$; and let $B_1 = U'\oplus I\ldots I\oplus V'''\oplus\ldots\oplus I,\ldots,B_{|K'''|} = U'\oplus I\oplus\ldots\oplus I\oplus V'''$. Here $V'' = U''$ or U''^*, according as $K'' > 0$ or $K'' < 0$, and similarly for V'''. $A_1A_2\ldots A_{|K''|}$ is an element of $\mathfrak{A}_1$ with symbol f'', and $B_1\ldots B_{|K'''|}$ is an element of $\mathfrak{A}_1$ with symbol f'''. The A_i's and B_j's are not themselves in $\mathfrak{A}_1$, but this causes no problem. If $\tilde{A}_i$ and $\tilde{B}_j$ are invertible trace class perturbations of A_i and B_j, then $\delta(f'',f''') = \Delta(\Pi\tilde{A}_i,\Pi\tilde{B}_j) = \Pi\Delta(\tilde{A}_i,\tilde{B}_j)$. Thus it is sufficient to prove $\Delta(\tilde{A}_i,\tilde{B}_j) = -1$, and the computation is the same for all values of i and j. There are still four cases involved, depending on the signs of K'' and K''', but the computation is exactly the same in all four cases. Thus the theorem has been reduced to the computation of a single determinant. This is very easy to do, assuming that the perturbations $\tilde{A}_i$ and $\tilde{B}_j$ are chosen in the obvious way. We find

$$\Delta(\tilde{A}_i,\tilde{B}_j) = \begin{vmatrix} 0 & 1 \\ 1 & 0 \end{vmatrix} = -1 . \qquad \text{Q.E.D.}$$

§7. Concluding Remarks

Helton and Howe made significant use of the formal properties of the trace invariant in establishing their formula for it. The major property was that $tr[p(X,Y),q(X,Y)] = 0$ if p and q are functions of a single polynomial. The analogous property for the determinant invariant is more subtle, and we therefore have not attempted to make our treatment of the determinant invariant fully parallel with Helton and Howe's treatment of the trace invariant. As an example of the subtlety, we point out that by the computation done above, there are two commuting partial isometries A and B, of index zero, such that if $\tilde{A}$ and $\tilde{B}$ are invertible operators which are trace class perturbations of A and B, then $\det(\tilde{A}\tilde{B}\tilde{A}^{-1}\tilde{B}^{-1}) = -1$.

Now, after the fact, we present what we think is a complete list, with some redundancy, of the formal properties of the determinant invariant:

i) $d(f,g) = d(g,f)^{-1}$

ii) $d(fg,h) = d(f,h) \cdot d(g,h)$

iii) $d(f,f) = (-1)^{\gamma(f)}$

iv) $d(f,c) = c^{\gamma(f)}$, c constant

v) $d(f,1-f) = 1$

vi) $d(f-a,f-b) = (-1)^{\gamma(f-b)}(a-b)^{\gamma(\frac{f-a}{f-b})}$, $a \neq b$.

Property v), which was suggested to us by H. Sah, is easy to prove if one observes that any smooth function f on X which omits the values 0 and 1 has a C^{∞} extension $\tilde{f}$ such that both $\tilde{f}$ and $1-\tilde{f}$ are admissible. iv) is obvious and vi) follows from iv) and v). iii), iv) and vi) enable one to compute $d(r(f),s(f))$, solely in terms of γ, for any rational functions r and s whose zeroes and poles are omitted by f. It will be noted that the functions f, g, etc. lie in the group of units of the ring of smooth functions on X, and therefore i) - vi) give an analogy with algebraic K-theory (cf.[6]). We think there is more to be said than is presently understood about the formal properties of the determinant invariant and the relationships with algebraic topology. It is difficult to formulate now what we think needs to be done because in the case considered here $(X \subset \mathbb{C})$ the algebraic topology is trivial. The higher dimensional case presents the real interest, and we hope to return to it later.

References

1. C. A. Berger and B. I. Shaw, "Self-commutators of multi-cyclic hyponormal operators are always trace class", to appear in Bull. A.M.S.

2. ———, "Intertwining, analytic structure, and the trace norm estimate, these Notes.

3. L. G. Brown, R. G. Douglas, and P. A. Fillmore, "Unitary equivalence modulo the compact operators and extensions of C*-algebras", these Notes.

4. J. W. Helton and R. E. Howe, "Commutators, traces, index and homology", these Notes.

5. J. G. Hocking and G. S. Young, Topology, Addison-Wesley, Reading, 1961 .

6. J. Milnor, "Algebraic K-theory and quadratic forms", Inventiones Math. 9 (1970) 318-344.

7. J. D. Pincus, "On the trace of commutators in the algebra of operators generated by an operator with trace-class self-commutator", to appear.

8. Richard W. Carey and J. D. Pincus, "An exponential formula for determining functions", to appear in Indiana Univ. Math. J.

S.U.N.Y. at Stony Brook
Stony Brook, New York

Vol. 215: P. Antonelli, D. Burghelea and P. J. Kahn, The Concordance-Homotopy Groups of Geometric Automorphism Groups. X, 140 pages. 1971. DM 16,–

Vol. 216: H. Maaß, Siegel's Modular Forms and Dirichlet Series. VII, 328 pages. 1971. DM 20,–

Vol. 217: T. J. Jech, Lectures in Set Theory with Particular Emphasis on the Method of Forcing. V, 137 pages. 1971. DM 16,–

Vol. 218: C. P. Schnorr, Zufälligkeit und Wahrscheinlichkeit. IV, 212 Seiten. 1971. DM 20,–

Vol. 219: N. L. Alling and N. Greenleaf, Foundations of the Theory of Klein Surfaces. IX, 117 pages. 1971. DM 16,–

Vol. 220: W. A. Coppel, Disconjugacy. V, 148 pages. 1971. DM 16,–

Vol. 221: P. Gabriel und F. Ulmer, Lokal präsentierbare Kategorien. V, 200 Seiten. 1971. DM 18,–

Vol. 222: C. Meghea, Compactification des Espaces Harmoniques. III, 108 pages. 1971. DM 16,–

Vol. 223: U. Felgner, Models of ZF-Set Theory. VI, 173 pages. 1971. DM 16,–

Vol. 224: Revètements Etales et Groupe Fondamental. (SGA 1). Dirigé par A. Grothendieck XXII, 447 pages. 1971. DM 30,–

Vol. 225: Théorie des Intersections et Théorème de Riemann-Roch. (SGA 6). Dirigé par P. Berthelot, A. Grothendieck et L. Illusie. XII, 700 pages. 1971. DM 40,–

Vol. 226: Seminar on Potential Theory, II. Edited by H. Bauer. IV, 170 pages. 1971. DM 18,–

Vol. 227: H. L. Montgomery, Topics in Multiplicative Number Theory. IX, 178 pages. 1971. DM 18,–

Vol. 228: Conference on Applications of Numerical Analysis. Edited by J. Ll. Morris. X, 358 pages. 1971. DM 26,–

Vol. 229: J. Väisälä, Lectures on n-Dimensional Quasiconformal Mappings. XIV, 144 pages. 1971. DM 16,–

Vol. 230: L. Waelbroeck, Topological Vector Spaces and Algebras. VII, 158 pages. 1971. DM 16,–

Vol. 231: H. Reiter, L^1-Algebras and Segal Algebras. XI, 113 pages. 1971. DM 16,–

Vol. 232: T. H. Ganelius, Tauberian Remainder Theorems. VI, 75 pages. 1971. DM 16,–

Vol. 233: C. P. Tsokos and W. J. Padgett. Random Integral Equations with Applications to stochastic Systems. VII, 174 pages. 1971. DM 18,–

Vol. 234: A. Andreotti and W. Stoll. Analytic and Algebraic Dependence of Meromorphic Functions. III, 390 pages. 1971. DM 26,–

Vol. 235: Global Differentiable Dynamics. Edited by O. Hájek, A. J. Lohwater, and R. McCann. X, 140 pages. 1971. DM 16,–

Vol. 236: M. Barr, P. A. Grillet, and D. H. van Osdol. Exact Categories and Categories of Sheaves. VII, 239 pages. 1971. DM 20,–

Vol. 237: B. Stenström, Rings and Modules of Quotients. VII, 136 pages. 1971. DM 16,–

Vol. 238: Der kanonische Modul eines Cohen-Macaulay-Rings. Herausgegeben von Jürgen Herzog und Ernst Kunz. VI, 103 Seiten. 1971. DM 16,–

Vol. 239: L. Illusie, Complexe Cotangent et Déformations I. XV, 355 pages. 1971. DM 26,–

Vol. 240: A. Kerber, Representations of Permutation Groups I. VII, 192 pages. 1971. DM 18,–

Vol. 241: S. Kaneyuki, Homogeneous Bounded Domains and Siegel Domains. V, 89 pages. 1971. DM 16,–

Vol. 242: R. R. Coifman et G. Weiss, Analyse Harmonique Non-Commutative sur Certains Espaces. V, 160 pages. 1971. DM 16,–

Vol. 243: Japan-United States Seminar on Ordinary Differential and Functional Equations. Edited by M. Urabe. VIII, 332 pages. 1971. DM 26,–

Vol. 244: Séminaire Bourbaki – vol. 1970/71. Exposés 382–399. IV, 356 pages. 1971. DM 26,–

Vol. 245: D. E. Cohen, Groups of Cohomological Dimension One. V, 99 pages. 1972. DM 16,–

Vol. 246: Lectures on Rings and Modules. Tulane University Ring and Operator Theory Year, 1970–1971. Volume I. X, 661 pages. 1972. DM 40,–

Vol. 247: Lectures on Operator Algebras. Tulane University Ring and Operator Theory Year, 1970–1971. Volume II. XI, 786 pages. 1972. DM 40,–

Vol. 248: Lectures on the Applications of Sheaves to Ring Theory. Tulane University Ring and Operator Theory Year, 1970–1971. Volume III. VIII, 315 pages. 1971. DM 26,–

Vol. 249: Symposium on Algebraic Topology. Edited by P. J. Hilton. VII, 111 pages. 1971. DM 16,–

Vol. 250: B. Jónsson, Topics in Universal Algebra. VI, 220 pages. 1972. DM 20,–

Vol. 251: The Theory of Arithmetic Functions. Edited by A. A. Gioia and D. L. Goldsmith VI, 287 pages. 1972. DM 24,–

Vol. 252: D. A. Stone, Stratified Polyhedra. IX, 193 pages. 1972. DM 18,–

Vol. 253: V. Komkov, Optimal Control Theory for the Damping of Vibrations of Simple Elastic Systems. V, 240 pages. 1972. DM 20,–

Vol. 254: C. U. Jensen, Les Foncteurs Dérivés de $\varprojlim$ et leurs Applications en Théorie des Modules. V, 103 pages. 1972. DM 16,–

Vol. 255: Conference in Mathematical Logic – London '70. Edited by W. Hodges. VIII, 351 pages. 1972. DM 26,–

Vol. 256: C. A. Berenstein and M. A. Dostal, Analytically Uniform Spaces and their Applications to Convolution Equations. VII, 130 pages. 1972. DM 16,–

Vol. 257: R. B. Holmes, A Course on Optimization and Best Approximation. VIII, 233 pages. 1972. DM 20,–

Vol. 258: Séminaire de Probabilités VI. Edited by P. A. Meyer. VI, 253 pages. 1972. DM 22,–

Vol. 259: N. Moulis, Structures de Fredholm sur les Variétés Hilbertiennes. V, 123 pages. 1972. DM 16,–

Vol. 260: R. Godement and H. Jacquet, Zeta Functions of Simple Algebras. IX, 188 pages. 1972. DM 18,–

Vol. 261: A. Guichardet, Symmetric Hilbert Spaces and Related Topics. V, 197 pages. 1972. DM 18,–

Vol. 262: H. G. Zimmer, Computational Problems, Methods, and Results in Algebraic Number Theory. V, 103 pages. 1972. DM 16,–

Vol. 263: T. Parthasarathy, Selection Theorems and their Applications. VII, 101 pages. 1972. DM 16,–

Vol. 264: W. Messing, The Crystals Associated to Barsotti-Tate Groups: With Applications to Abelian Schemes. III, 190 pages. 1972. DM 18,–

Vol. 265: N. Saavedra Rivano, Catégories Tannakiennes. II, 418 pages. 1972. DM 26,–

Vol. 266: Conference on Harmonic Analysis. Edited by D. Gulick and R. L. Lipsman. VI, 323 pages. 1972. DM 24,–

Vol. 267: Numerische Lösung nichtlinearer partieller Differential- und Integro-Differentialgleichungen. Herausgegeben von R. Ansorge und W. Törnig, VI, 339 Seiten. 1972. DM 26,–

Vol. 268: C. G. Simader, On Dirichlet's Boundary Value Problem. IV, 238 pages. 1972. DM 20,–

Vol. 269: Théorie des Topos et Cohomologie Etale des Schémas. (SGA 4). Dirigé par M. Artin, A. Grothendieck et J. L. Verdier. XIX, 525 pages. 1972. DM 50,–

Vol. 270: Théorie des Topos et Cohomologie Etale des Schémas. Tome 2. (SGA 4). Dirigé par M. Artin, A. Grothendieck et J. L. Verdier. V, 418 pages. 1972. DM 50,–

Vol. 271: J. P. May, The Geometry of Iterated Loop Spaces. IX, 175 pages. 1972. DM 18,–

Vol. 272: K. R. Parthasarathy and K. Schmidt, Positive Definite Kernels, Continuous Tensor Products, and Central Limit Theorems of Probability Theory. VI, 107 pages. 1972. DM 16,–

Vol. 273: U. Seip, Kompakt erzeugte Vektorräume und Analysis. IX, 119 Seiten. 1972. DM 16,–

Vol. 274: Toposes, Algebraic Geometry and Logic. Edited by. F. W. Lawvere. VI, 189 pages. 1972. DM 18,–

Vol. 275: Séminaire Pierre Lelong (Analyse) Année 1970–1971. VI, 181 pages. 1972. DM 18,–

Vol. 276: A. Borel, Représentations de Groupes Localement Compacts. V, 98 pages. 1972. DM 16,–

Vol. 277: Séminaire Banach. Edité par C. Houzel. VII, 229 pages. 1972. DM 20,–

Vol. 278: H. Jacquet, Automorphic Forms on GL(2). Part II. XIII, 142 pages. 1972. DM 16,–

Vol. 279: R. Bott, S. Gitler and I. M. James, Lectures on Algebraic and Differential Topology. V, 174 pages. 1972. DM 18,–

Vol. 280: Conference on the Theory of Ordinary and Partial Differential Equations. Edited by W. N. Everitt and B. D. Sleeman. XV, 367 pages. 1972. DM 26,–

Vol. 281: Coherence in Categories. Edited by S. Mac Lane. VII, 235 pages. 1972. DM 20,–

Vol. 282: W. Klingenberg und P. Flaschel, Riemannsche Hilbertmannigfaltigkeiten. Periodische Geodätische. VII, 211 Seiten. 1972. DM 20,–

Vol. 283: L. Illusie, Complexe Cotangent et Déformations II. VII, 304 pages. 1972. DM 24,–

Vol. 284: P. A. Meyer, Martingales and Stochastic Integrals I. VI, 89 pages. 1972. DM 16,–

Vol. 285: P. de la Harpe, Classical Banach-Lie Algebras and Banach-Lie Groups of Operators in Hilbert Space. III, 160 pages. 1972. DM 16,–

Vol. 286: S. Murakami, On Automorphisms of Siegel Domains. V, 95 pages. 1972. DM 16,–

Vol. 287: Hyperfunctions and Pseudo-Differential Equations. Edited by H. Komatsu. VII, 529 pages. 1973. DM 36,–

Vol. 288: Groupes de Monodromie en Géométrie Algébrique. (SGA 7 I). Dirigé par A. Grothendieck. IX, 523 pages. 1972. DM 50,–

Vol. 289: B. Fuglede, Finely Harmonic Functions. III, 188. 1972. DM 18,–

Vol. 290: D. B. Zagier, Equivariant Pontrjagin Classes and Applications to Orbit Spaces. IX, 130 pages. 1972. DM 16,–

Vol. 291: P. Orlik, Seifert Manifolds. VIII, 155 pages. 1972. DM 16,–

Vol. 292: W. D. Wallis, A. P. Street and J. S. Wallis, Combinatorics: Room Squares, Sum-Free Sets, Hadamard Matrices. V, 508 pages. 1972. DM 50,–

Vol. 293: R. A. DeVore, The Approximation of Continuous Functions by Positive Linear Operators. VIII, 289 pages. 1972. DM 24,–

Vol. 294: Stability of Stochastic Dynamical Systems. Edited by R. F. Curtain. IX, 332 pages. 1972. DM 26,–

Vol. 295: C. Dellacherie, Ensembles Analytiques, Capacités, Mesures de Hausdorff. XII, 123 pages. 1972. DM 16,–

Vol. 296: Probability and Information Theory II. Edited by M. Behara, K. Krickeberg and J. Wolfowitz. V, 223 pages. 1973. DM 20,–

Vol. 297: J. Garnett, Analytic Capacity and Measure. IV, 138 pages. 1972. DM 16,–

Vol. 298: Proceedings of the Second Conference on Compact Transformation Groups. Part 1. XIII, 453 pages. 1972. DM 32,–

Vol. 299: Proceedings of the Second Conference on Compact Transformation Groups. Part 2. XIV, 327 pages. 1972. DM 26,–

Vol. 300: P. Eymard, Moyennes Invariantes et Représentations Unitaires. II. 113 pages. 1972. DM 16,–

Vol. 301: F. Pittnauer, Vorlesungen über asymptotische Reihen. VI, 186 Seiten. 1972. DM 18,–

Vol. 302: M. Demazure, Lectures on p-Divisible Groups. V, 98 pages. 1972. DM 16,–

Vol. 303: Graph Theory and Applications. Edited by Y. Alavi, D. R. Lick and A. T. White. IX, 329 pages. 1972. DM 26,–

Vol. 304: A. K. Bousfield and D. M. Kan, Homotopy Limits, Completions and Localizations. V, 348 pages. 1972. DM 26,–

Vol. 305: Théorie des Topos et Cohomologie Etale des Schémas. Tome 3. (SGA 4). Dirigé par M. Artin, A. Grothendieck et J. L. Verdier. VI, 640 pages. 1973. DM 50,–

Vol. 306: H. Luckhardt, Extensional Gödel Functional Interpretation. VI, 161 pages. 1973. DM 18,–

Vol. 307: J. L. Bretagnolle, S. D. Chatterji et P.-A. Meyer, Ecole d'été de Probabilités: Processus Stochastiques. VI, 198 pages. 1973. DM 20,–

Vol. 308: D. Knutson, λ-Rings and the Representation Theory of the Symmetric Group. IV, 203 pages. 1973. DM 20,–

Vol. 309: D. H. Sattinger, Topics in Stability and Bifurcation Theory. VI, 190 pages. 1973. DM 18,–

Vol. 310: B. Iversen, Generic Local Structure of the Morphisms in Commutative Algebra. IV, 108 pages. 1973. DM 16,–

Vol. 311: Conference on Commutative Algebra. Edited by J. W. Brewer and E. A. Rutter. VII, 251 pages. 1973. DM 22,–

Vol. 312: Symposium on Ordinary Differential Equations. Edited by W. A. Harris, Jr. and Y. Sibuya. VIII, 204 pages. 1973. DM 22,–

Vol. 313: K. Jörgens and J. Weidmann, Spectral Properties of Hamiltonian Operators. III, 140 pages. 1973. DM 16,–

Vol. 314: M. Deuring, Lectures on the Theory of Algebraic Functions of One Variable. VI, 151 pages. 1973. DM 16,–

Vol. 315: K. Bichteler, Integration Theory (with Special Attention to Vector Measures). VI, 357 pages. 1973. DM 26,–

Vol. 316: Symposium on Non-Well-Posed Problems and Logarithmic Convexity. Edited by R. J. Knops. V, 176 pages. 1973. DM 18,–

Vol. 317: Séminaire Bourbaki – vol. 1971/72. Exposés 400–417. IV, 361 pages. 1973. DM 26,–

Vol. 318: Recent Advances in Topological Dynamics. Edited by A. Beck, VIII, 285 pages. 1973. DM 24,–

Vol. 319: Conference on Group Theory. Edited by R. W. Gatterdam and K. W. Weston. V, 188 pages. 1973. DM 18,–

Vol. 320: Modular Functions of One Variable I. Edited by W. Kuyk. V, 195 pages. 1973. DM 18,–

Vol. 321: Séminaire de Probabilités VII. Edité par P. A. Meyer. VI, 322 pages. 1973. DM 26,–

Vol. 322: Nonlinear Problems in the Physical Sciences and Biology. Edited by I. Stakgold, D. D. Joseph and D. H. Sattinger. VIII, 357 pages. 1973. DM 26,–

Vol. 323: J. L. Lions, Perturbations Singulières dans les Problèmes aux Limites et en Contrôle Optimal. XII, 645 pages. 1973. DM 42,–

Vol. 324: K. Kreith, Oscillation Theory. VI, 109 pages. 1973. DM 16,–

Vol. 325: Ch.-Ch. Chou, La Transformation de Fourier Complexe et L'Equation de Convolution. IX, 137 pages. 1973. DM 16,–

Vol. 326: A. Robert, Elliptic Curves. VIII, 264 pages. 1973. DM 22,–

Vol. 327: E. Matlis, 1-Dimensional Cohen-Macaulay Rings. XII, 157 pages. 1973. DM 18,–

Vol. 328: J. R. Büchi and D. Siefkes, The Monadic Second Order Theory of All Countable Ordinals. VI, 217 pages. 1973. DM 20,–

Vol. 329: W. Trebels, Multipliers for (C, α)-Bounded Fourier Expansions in Banach Spaces and Approximation Theory. VII, 103 pages. 1973. DM 16,–

Vol. 330: Proceedings of the Second Japan-USSR Symposium on Probability Theory. Edited by G. Maruyama and Yu. V. Prokhorov. VI, 550 pages. 1973. DM 36,–

Vol. 331: Summer School on Topological Vector Spaces. Edited by L. Waelbroeck. VI, 226 pages. 1973. DM 20,–

Vol. 332: Séminaire Pierre Lelong (Analyse) Année 1971-1972. V, 131 pages. 1973. DM 16,–

Vol. 333: Numerische, insbesondere approximationstheoretische Behandlung von Funktionalgleichungen. Herausgegeben von R. Ansorge und W. Törnig. VI, 296 Seiten. 1973. DM 24,–

Vol. 334: F. Schweiger, The Metrical Theory of Jacobi-Perron Algorithm. V, 111 pages. 1973. DM 16,–

Vol. 335: H. Huck, R. Roitzsch, U. Simon, W. Vortisch, R. Walden, B. Wegner und W. Wendland, Beweismethoden der Differentialgeometrie im Großen. IX, 159 Seiten. 1973. DM 18,–

Vol. 336: L'Analyse Harmonique dans le Domaine Complexe. Edité par E. J. Akutowicz. VIII, 169 pages. 1973. DM 18,–

Vol. 337: Cambridge Summer School in Mathematical Logic. Edited by A. R. D. Mathias and H. Rogers. IX, 660 pages. 1973. DM 42,–

Vol: 338: J. Lindenstrauss and L. Tzafriri, Classical Banach Spaces. IX, 243 pages. 1973. DM 22,–

Vol. 339: G. Kempf, F. Knudsen, D. Mumford and B. Saint-Donat, Toroidal Embeddings I. VIII, 209 pages. 1973. DM 20,–

Vol. 340: Groupes de Monodromie en Géométrie Algébrique. (SGA 7 II). Par P. Deligne et N. Katz. X, 438 pages. 1973. DM 40,–

Vol. 341: Algebraic K-Theory I, Higher K-Theories. Edited by H. Bass. XV, 335 pages. 1973. DM 26,–

Vol. 342: Algebraic K-Theory II, "Classical" Algebraic K-Theory, and Connections with Arithmetic. Edited by H. Bass. XV, 527 pages. 1973. DM 36,–